GENETIC ALGORITHMS FOR MACHINE LEARNING

edited by

John J. Grefenstette
Naval Research Laboratory

A Special Issue of MACHINE LEARNING

Reprinted from
MACHINE LEARNING
Vol. 13, Nos. 2- 3 (1993)

Kluwer Academic Publishers
Boston/Dordrecht/London

CONTENTS

Distributors for North America:
Kluwer Academic Publishers
101 Philip Drive
Assinippi Park
Norwell, Massachusetts 02061 USA

Distributors for all other countries:
Kluwer Academic Publishers Group
Distribution Centre
Post Office Box 322
3300 AH Dordrecht, THE NETHERLANDS

Library of Congress Cataloging-in-Publication Data

Genetic algorithms for machine learning / John J. Grefenstette, editor.
p. cm.
"A Special issue of Machine learning."
"Reprinted from Machine learning, vol. 13, nos. 2-3 (1993)."
Includes index.
ISBN 0-7923-9407-0
1. Machine learning. 2. Genetic algorithms. I. Grefenstette, John J. II. Machine learning. Special issue.
Q325.5.G45 1994
006.3'1--dc20 93-34282
CIP

Printed on acid-free paper.

Printed in the United States of America

Machine Learning, 13, 157–160 (1993)

Introduction

It is my pleasure to introduce this third Special Issue on Genetic Algorithms (GAs). The articles presented here were selected from preliminary versions presented at the International Conference on Genetic Algorithms in June 1991, as well as at a special Workshop on Genetic Algorithms for Machine Learning at the same Conference.

Genetic algorithms are general-purpose search algorithms that use principles inspired by natural population genetics to evolve solutions to problems (Holland, 1975). The basic idea is to maintain a population of knowledge structures that represent candidate solutions to the problem of interest. The population evolves over time through a process of competition (i.e., survival of the fittest) and controlled variation (i.e., recombination and mutation). For further background material on GAs for machine learning, interested readers are referred to Davis (1991), De Jong (1990), Goldberg (1989), and Holland (1986).

Previous special issues on GAs (volume 3, numbers 2–3; volume 5, number 4) as well as articles appearing in regular issues of *Machine Learning*, have given a fair indication of the rich variety of work in the field. Although there certainly has not been exhaustive coverage of any topic, there have been several good illustrations of the use of genetic algorithms in classifier systems, in systems that learn sequential decision rules, and in systems that perform parameter learning. Taking the previous distribution of topics into account, the current issue is devoted to filling in some of the gaps. Consequently, this issue contains articles on three topics that have not been the focus of many preivous articles on GAs in *Machine Learning*, namely, concept learning from examples, reinforcement learning for control, and theoretical analysis of GAs. It is hoped that this sample will serve to broaden the acquaintance of the general machine learning community with the major areas of work on GAs.

The articles in this issue address a number of central issues in applying GAs to machine learning problems. For example, the choice of an appropriate representation and the corresponding set of genetic learning operators is an important set of decisions facing a user of a genetic algorithm. The first two articles in this issue illustrate two contrasting approaches. One approach is to adopt the traditional string representation and its corresponding crossover and mutation operators. This approach offers the advantage of a good fit with the theoretical analysis that has been performed to date, as well as the convenience of using widely available software. The article by De Jong, Spears, and Gordon takes this minimalist approach to the problem of learning concepts from examples. A second approach is taken in the article by Janikow. The idea here is to adopt a representation (e.g., symbolic rules) that more closely reflects that semantics of the specific problem to be solved (e.g., concept learning), and to design specialized genetic operators (e.g., rule specialization, rule generalization) that take advantage of the particular representation being used. The second approach offers the possibility of exploiting knowledge about the problem being solved in the hope of achieving a more efficient search. The second approach also offers the advantage of possibly combining genetic algorithms with other problem-solving methods, as illustrated by the SAMUEL system in the domain of sequential decison making (Grefenstette, Ramsey, &

Schultz, 1990), but it relies on an assumption that the important search properties of GAs carry over to the new formulation. The results presented here show that both approaches can work reasonably well, compared to traditional symbolic concept learning systems. Whether one approach to GA-based learning should be preferred over the other remains an interesting open question.

Both of the first two articles also explore the issue of dynamically adjusting the bias of the genetic search during learning. De Jong et al. adjust the bias explicitly by including switches for alternative operators as part of the evolving knowledge structures. Janikow dynamically adjusts the probability of applying his extended set of operators based on the consistency and completeness of the current hypotheses. Both articles reflect a growing trend in GAs to reduce the number of free parameters in a GA by enabling the system to adapt its internal operations during the course of the search.

Another important dimension along with GAs differ is whether the knowledge structures in the population represent independent solutions or dependent components of a single solution. For example, each structure in SAMUEL represents an entire set of decision rules (Grefenstette et al., 1990), whereas the individuals in classifier systems represents single rules (Holland, 1986; Wilson 1987). The first two articles in this issue take the former approach, while the third article in this issue, by Greene and Smith, adopts the latter approach to concept learning. The innovation here is to measure the fitness of each rule in terms of both its classification accuracy and its ability to cover training examples not covered by other rules. Empirical results show that this "population-as-model" approach is competitive with existing symbolic induction methods.

The first three articles use what is commonly called the "generational model" of GAs. In this model, during each generation, a new population of knowledge structures is formed in two steps. First, structures in the current population are selected to be reproduced on the basis of their relative fitness. That is, high-performing structures may be chosen several times for replication, and poorly performing structures may not be chosen at all. Second, the selected structures are recombined using idealized genetic operators to form a new set of structures that are then evaluated as solutions to the given problem. The fourth article in this issue, by Whitley, Dominic, Das, and Anderson, presents a different model, called the "steady-state" GA. In the steady-state model, the GA selects two parent structures, recombines them, evaluates them, and adds them to the population, replacing some older structures. These two approaches represent different ways to simulate an essentially parallel algorithm on a sequential computer The mathematical differences between the approaches depend on the details of the selection algorithms, and are still the object of discussion within the GA community. The steady-state approach appears to offer certain implementation advantages on massively parallel architectures. Whitley et al. apply the steady-state approach to a classic problem of learning to control an inverted pendulum system, using the GA to learn weights for a neural network. A set of empirical comparisons between the GA and the well-known Adaptive Heuristic Critic (AHC) show that rates of performance improvements are comparable, but the GA appears to be more robust in the sense that it can learn to successfully control the system from a wider range of initial conditions.

The reliance on empirical studies in the preceding articles is an indication of the current state of the art in GAs: we are clearly still in an experimental stage. Nonetheless, ever since Holland's original analysis, there has always been a high level of interest in developing a more theoretical characterization of the information processing mechanisms at work in

GAs. Viewed in the most general light, GAs gather information about regularities in the search space defined by the problem representation and the fitness function, and dynamically allocate future search effort to promising regions of that space. Some central questions are, then, what classes of landscapes are most easy searched by GAs, and what classes present the most challenging problems for GAs. While definitive answers are yet to be found, a number of interesting analytic tools have been developed that may help to characterize the difficulities associated with different fitness functions. The final article in this issue, by Forrest and Mitchell, provides a tutorial on some of these analytic methods for GAs. The article also provides an admirable example of replicating some previously reported empirical results that at first appeared puzzling, and then performing additional studies to distinguish among alternative theoretical explanations.

Beyond the results appearing in these articles, there have been a number of other interesting recent developments within the GA field. In the last few years, there has been an increasing level of communication and cooperation between formerly independent research communities. In North America, most of the GA community has been elaborating on the framework developed by John Holland during the 1960s and 1970s at the University of Michigan. In Europe, a similar framework, called the *EvolutionStrategie* (ES), was developed independently about the same time by Rechenberg and Schwefel (see back, Hoffmeister, & H.-P. Schwefel, 1991, for a survey of this area). While the GA community has focused on simple string representations that support general-purpose problem solving, and binary recombination operators like crossover, the ES community has generally focused on engineering applications involving function optimization, relying on sophisticated mutation operators to create variants in the population. This work has reawakened interest in the early work on mutation-based Evolutionary Programming (Fogel, Owens, & Walsh, 1966). The cross-fertilzation of these research communities has begun to bear fruit (Schwefel & Maenner, 1991), and there is growing interest in moving toward a common framework, which has been called *Evolutionary Computation.*

The theoretical foundations of GAs continues to receive a high level of attention. Besides sessions devoted to theory within the main Conferences, there have been two specialized workshops on Foundation of Genetic Algorithms (FOGA), focusing on such issues as the mathematical models of genetic algorithms, characterization of problem difficulty, the role of representation, and the effects of various forms of genetic operators. The article by Forrest and Mitchell gives a sampling of the flavor of this work. For more of these topics, interested readers are referred to edited versions of the FOGA proceedings, which have been published in book form (Rawlins, 1991; Whitley, 1992).

Applications are also proceeding along a wide range of domains. A good survey of successful applications, along with a tutorial on implementation strategies, appears in the *Handbook of Genetic Algorithms*, edited by L. Davis (1991). In summary, the study of genetic algorithms is proceeding at a robust pace. If experimental progress and theoretical understanding continue to evolve as expected, genetic algorithms will continue to provide a distinctive approach to machine learning.

John J. Grefenstette
Navy Center for Applied Research in AI
Naval Research Laboratory
Washington, DC 20375-5000

References

Back, T., Hoffmeister, F., & Schwefel, H.-P. (1991). A survey of evolution strategies. *Proceedings of the Fourth International Conference on Genetic Algorithms* (pp. 2–9). San Mateo, CA: Morgan Kaufmann.

Davis, L. (Ed.). (1991). *Handbook of genetic algorithms*. New York: Van Nostrand Reinhold.

De Jong, K.A. (1990). Genetic-algorithm-based learning. In Y. Kodratoff & R. Michalski (Eds.), *Machine learning: An artifical intelligence approach* (Vol. 3). San Mateo, CA: Morgan Kaufmann.

Fogel, L.J., Owens, A.J., & Walsh, M.J. (1966). *Artificial intelligence through simulated evolution*. New York: Wiley.

Goldberg, D.E. (1989). *Genetic algorithms in search, optimization, and machine learning*. Reading, MA: Addison-Wesley.

Grefenstette, J.J., Ramsey, C.L., & Schultz, A.C. (1990). Learning sequential decision rules using simulation models and competition. *Machine Learning*, 5(4), 335–381.

Holland, J.H. (1975). *Adaptation in natural and artificial systems*. Cambridge, MA: MIT Press.

Holland, J.H. (1986). Escaping brittleness: The possibilities of general-purpose learning algorithms applied to parallel rule-based systems. In R.S. Michalski, J.G. Carbonnell, & T.M. Mitchell (Eds.), *Machine learning: An artificial intelligence approach* (Vol. 2). Los Altos, CA: Morgan Kaufmann

Rawlins, G.J.E. (Ed.). (1991). *Foundations of genetic algorithms*. San Mateo, CA: Morgan Kaufmann.

Schwefel, H.-P., & R. Maenner (1991). *Parallel problem solving from nature*. Lecture Notes in Computer Science 496. Berlin: Springer-Verlag.

Whitley, D. (Ed.). (1992). *Foundations of genetic algorithms 2*. San Mateo, CA: Morgan Kaufmann.

Wilson, S.W. (1987). Classifier systems and the animat problem. *Machine Learning*, 2, 199–228.

Machine Learning, 13, 161-188 (1993)

Using Genetic Algorithms for Concept Learning

KENNETH A. DE JONG KDEJONG@CS.GMU.EDU
Computer Science Department, George Mason University, Fairfax, VA 22030

WILLIAM M. SPEARS SPEARS@AIC.NRL.NAVY.MIL
Naval Research Laboratory, Code 5510, Washington, DC 20375

DIANA F. GORDON GORDON@AIC.NRL.NAVY.MIL
Naval Research Laboratory, Code 5510, Washington, DC 20375

Abstract. In this article, we explore the use of genetic algorithms (GAs) as a key element in the design and implementation of robust concept learning systems. We describe and evaluate a GA-based system called GABIL that continually learns and refines concept classification rules from its interaction with the environment. The use of GAs is motivated by recent studies showing the effects of various forms of bias built into different concept learning systems, resulting in systems that perform well on certain concept classes (generally, those well matched to the biases) and poorly on others. By incorporating a GA as the underlying adaptive search mechanism, we are able to construct a concept learning system that has a simple, unified architecture with several important features. First, the system is surprisingly robust even with minimal bias. Second, the system can be easily extended to incorporate traditional forms of bias found in other concept learning systems. Finally, the architecture of the system encourages explicit representation of such biases and, as a result, provides for an important additional feature: the ability to *dynamically* adjust system bias. The viability of this approach is illustrated by comparing the performance of GABIL with that of four other more traditional concept learners (AQ14, C4.5, ID5R, and IACL) on a variety of target concepts. We conclude with some observations about the merits of this approach and about possible extensions.

Keywords. Concept learning, genetic algorithms, bias adjustment

1. Introduction

An important requirement for both natural and artificial organisms is the ability to acquire concept classification rules from interactions with their environment. In this article, we explore the use of an adaptive search technique, namely, genetic algorithms (GAs), as the central mechanism for designing such systems. The motivation for this approach comes from an accumulating body of evidence that suggests that, although concept learners require fairly strong biases to induce classification rules efficiently, no a priori set of biases is appropriate for all concept learning tasks. We have been exploring the design and implementation of more robust concept learning systems that are capable of adaptively shifting their biases when appropriate. What we find particularly intriguing is the natural way GA-based concept learners can provide this capability.

As proof of concept we have implemented a system called GABIL with these features and have compared its performance with four more traditional concept learning systems (AQ14, C4.5, ID5R, and IACL) on a set of target concepts of varying complexity.

We present these results in the following manner. We begin by showing how concept learning tasks can be represented and solved by traditional GAs with minimal implicit bias. We illustrate this by explaining the GABIL system architecture in some detail.

We then compare the performance of this minimalist GABIL system with AQ14, C4.5, ID5R, and IACL on a set of target concepts of varying complexity. As expected, no single system is best for all of the presented concepts. However, a posteriori, one can identify the biases that were largely responsible for each system's superiority on certain classes of target concepts.

We then show how GABIL can be easily extended to include these biases, which improves system performance on various classes of concepts. However, the introduction of additional biases raises the problem of how and when to apply them to achieve the bias adjustment necessary for more robust performance.

Finally, we show how a GA-based system can be extended to dynamically adjust its own bias in a very natural way. We support these claims with empirical studies showing the improvement in robustness of GABIL with adaptive bias, and we conclude with a discussion of the merits of this approach and directions for further research.

2. GAs and concept learning

Supervised concept learning involves inducing descriptions (i.e., inductive hypotheses) for the concepts to be learned from a set of positive and negative examples of the target concepts. Examples (instances) are represented as points in an n-dimensional feature space that is defined a priori and for which all the legal values of the features are known. Concepts are therefore represented as subsets of points in the given n-dimensional space. A concept learning program is presented with both a description of the feature space and a set of correctly classified examples of the concepts, and is expected to generate a reasonably accurate description of the (unknown) concepts.

The choice of the concept description language is important in several respects. It introduces a language bias that can make some classes of concepts easy to describe while other class descriptions become awkward and difficult. Most of the approaches have involved the use of classification rules, decision trees, or, more recently, neural networks. Each such choice also defines a space of all possible concept descriptions from which the "correct" concept description must be selected using a given set of positive and negative examples as constraints. In each case the size and complexity of this search space requires fairly strong additional heuristic pruning in the form of biases towards concepts that are "simpler," "more general," and so on.

The effects of adding such biases in addition to the language bias is to produce systems that work well on concepts that are well matched to these biases, but perform poorly on other classes of concepts. What is needed is a means for improving the overall robustness and adaptability of these concept learners in order to successfully apply them to situations in which little is known a priori about the concepts to be learned. Since genetic algorithms (GAs) have been shown to be a powerful adaptive search technique for large, complex spaces in other contexts, our motivation for this work is to explore their usefulness in building more flexible and effective concept learners.[1]

In order to apply GAs to a concept learning problem, we need to select an internal representation of the space to be searched. This must be done carefully to preserve the properties that make the GAs effective adaptive search procedures (see DeJong (1987) for a more detailed discussion). The traditional internal representation of GAs involves using fixed-length (generally binary) strings to represent points in the space to be searched. However, such representations do not appear well suited for representing the space of concept descriptions that are generally symbolic in nature, that have both syntactic and semantic constraints, and that can be of widely varying length and complexity.

There are two general approaches one might take to resolve this issue. The first involves changing the fundamental GA operators (crossover and mutation) to work effectively with complex non-string objects. Alternatively, one can attempt to construct a string representation that minimizes any changes to the GA. Each approach has certain advantages and disadvantages. Developing new GA operators that are sensitive to the syntax and semantics of symbolic concept descriptions is appealing and can be quite effective, but also introduces a new set of issues relating to the precise form such operators should take and the frequency with which they should be applied. The alternative approach, using a string representation, puts the burden on the system designer to find a mapping of complex concept descriptions into linear strings that has the property that the traditional GA operators that manipulate these strings preserve the syntax and semantics of the underlying concept descriptions. The advantage of this approach is that, if an effective mapping can be defined, a standard "off the shelf" GA can be used with few, if any, changes. In this article, we illustrate the latter approach and develop a system that uses a traditional GA with minimal changes. For examples of the other approach, see Rendell (1985), Grefenstette (1989), Koza (1991), and Janikow (1991).

The decision to adopt a minimalist approach has immediate implications for the choice of concept description languages. We need to identify a language that can be effectively mapped into string representations and yet retains the necessary expressive power to represent complex descriptions efficiently. As a consequence, we have chosen a simple, yet general rule language for describing concepts, the details of which are presented in the following sections.

2.1. Representing the search space

A natural way to express complex concepts is as a disjunctive set of possibly overlapping classification rules, i.e., in disjunctive normal form (DNF). The left-hand side of each rule (i.e., disjunct or term) consists of a conjunction of one or more tests involving feature values. The right-hand side of a rule indicates the concept (classification) to be assigned to the examples that are matched (covered) by the left-hand side of the rule. Collectively, a set of such rules can be thought of as representing the unknown concept if the rules correctly classify the elements of the feature space.

If we allow arbitrarily complex terms in the conjunctive left-hand side of such rules, we will have a very powerful description language that will be difficult to represent as strings. However, by restricting the complexity of the elements of the conjunctions, we are able to use a string representation and standard GAs, with the only negative side effect that more rules may be required to express the concept. This is achieved by restricting each element of a conjunction to be a test of the form:

> If the value of feature *i* of the example is in the given value set, return true else, return false.

For example, a rule might take the following symbolic form:

> If (Fl = large) and (F2 = sphere or cube) then it is a widget.

Since the left-hand sides are conjunctive forms with internal disjunction (e.g., the disjunction within feature F2), there is no loss of generality by requiring that there be at most one test for each feature (on the left-hand side of a rule). The result is a modified DNF that allows internal disjunction. (See Michalski (1983) for a discussion of internal disjunction.)

With these restrictions we can now construct a fixed-length internal representation for classification rules. Each fixed-length rule will have *N* feature tests, one for each feature. Each feature test will be represented by a fixed-length binary string, the length of which will depend on the type of feature (nominal, ordered, etc.). Currently, GABIL only uses features with nominal values. The system uses *k* bits for the *k* values of a nominal feature. So, for example, if the set of legal values for feature Fl is {*small*, *medium*, *large*}, then the pattern 011 would represent the test for Fl being *medium* or *large*.

Further suppose that feature F2 has the values {*sphere*, *cube*, *brick*, *tube*} and there are two classes, *widgets* and *gadgets*. Then, a rule for this two-feature problem would be represented internally as

Fl	F2	Class
111	1000	0

This rule is equivalent to

> If (Fl = small or medium or large) and (F2 = sphere) then it is a widget.

Notice that a feature test involving all 1's matches any value of a feature and is equivalent to "dropping" that conjunctive term (i.e., the feature is irrelevant for that rule). So, in the example above, only the values of F2 are relevant, and the rule is more succinctly interpreted as

> If (F2 = sphere) then it is a widget.

For completeness, we allow patterns of all 0's, which match nothing. This means that any rule containing such a pattern will not match any points in the feature space. While rules of this form are of no use in the final concept description, they are quite useful as storage areas for GAs when evolving and testing sets of rules.

The right-hand side of a rule is simply the class (concept) to which the example belongs. This means that our rule language defines a "stimulus-response" system with no message passing or any other form of internal memory such as those found in Holland (1986). In many of the traditional concept learning contexts, there is only a single concept to be learned. In these situations there is no need for the rules to have an explicit right-hand side, since the class is implied. Clearly, the string representation we have chosen handles such cases easily by assigning no bits for the right-hand side of each rule.

2.1.1. Sets of classification rules

Since a concept description will consist of one or more classification rules, we still need to specify how GAs will be used to evolve sets of rules. There are currently two basic strategies: the Michigan approach, exemplified by Holland's classifier system (Holland, 1986), and the Pittsburgh approach, exemplified by Smith's LS-1 system (Smith, 1983). Systems using the Michigan approach maintain a population of *individual rules* that compete with each other for space and priority in the population. In contrast, systems using the Pittsburgh approach maintain a population of *variable-length rule sets* that compete with each other with respect to performance on the domain task. There is still much to be learned about the relative merits of the two approaches. In this article we report on results obtained from using the Pittsburgh approach.[2] That is, each individual in the population is a variable-length string representing an unordered set of fixed-length rules. The number of rules in a particular individual can be unrestricted or limited by a user-defined upper bound.

To illustrate this representation more concretely, consider the following example of a rule set with two rules:

F1	F2	Class	F1	F2	Class
100	1111	0	011	0010	0

This rule set is equivalent to

If (F1 = small) then it is a widget
or
If ((F1 = medium or large) and (F2 = brick)) then it is a widget.

2.1.2. Rule set execution semantics

In choosing a rule set representation for use with GAs, it is also important to define simple execution semantics that encourage the development of rule subsets and their subsequent recombination with other subsets to form new and better rule sets. One important feature of our execution semantics with this property is that there is no order-dependency among the rules in a rule set. When a rule set is used to predict the class of an example, the left-hand sides of all rules in a rule set are checked to see if they match (cover) a particular example. This "parallel" execution semantics means that rules perform in a location-independent manner.

It is possible that an example might be covered by more than one rule. There are a number of existing approaches for resolving such conflicts on the basis of dynamically calculated rule strengths, by measuring the complexity of the left-hand sides of rules, or by various voting schemes. It is also possible that there are no rules that cover a particular example. Unmatched examples could be handled by partial matching and/or covering operators. How best to handle these two situations in the general context of learning multiple concepts (classes) simultaneously is a difficult issue that we have not yet resolved to our satisfaction.

However, these issues are considerably simpler when single concepts are being learned. In this case, since there is only one class, the right-hand sides of all rules are the same and do not need to be explicitly represented. Hence, it is quite natural to view the rules in a rule set as a union of (possibly overlapping) covers of the concept to be learned. An example that matches one or more rules is classified as a positive example of the concept, and an example that fails to match any rule is classified as a negative example.

In this article, we focus on the simpler case of single-concept learning problems (which have also dominated the concept-learning literature). We have left the extension to multi-concept problems for future work.

2.1.3. Crossover and mutation

Genetic operators modify individuals within a population to produce new individuals for testing and evaluation. Historically, crossover and mutation have been the most important and best understood genetic operators. Crossover takes two individuals and produces two new individuals, by swapping portions of genetic material (e.g., bits). Mutation simply flips random bits within the population, with a small probability (e.g., 1 bit per 1000). One of our goals was to achieve a concept learning representation that could exploit these fundamental operators. We feel we have achieved that goal with the variable-length string representation involving fixed-length rules described in the previous sections.

Our mutation operator is identical to the standard one and performs the usual bit-level mutations. We are currently using a fairly standard extension of the traditional two-point crossover operator in order to handle variable-length rule sets.[3] With standard two-point crossover on fixed-length strings, there are only two degrees of freedom in selecting the crossover points, since the crossover points always match up on both parents (e.g., exchanging the segments from positions 12–25 on each parent). However, with variable length strings there are four degrees of freedom, since there is no guarantee that, having picked two crossover points for the first parent, the same points exist on the second parent. Hence, a second set of crossover points must be selected for it.

As with standard crossover, there are no restrictions on where the crossover points may occur (i.e., both on rule boundaries and within rules). The only requirement is that the corresponding crossover points on the two parents "match up semantically." That is, if one parent is being cut on a rule boundary, then the other parent must be cut on a rule boundary. Similarly, if one parent is being cut at a point 5 bits to the right of a rule boundary, then the other parent must be cut in a similar spot (i.e., 5 bits to the right of some rule boundary). As an example, consider the following two rule sets:

F1	F2	Class	F1	F2	Class
100	**0100**	**0**	**011**	**0010**	**0**
010	0001	0	110	0011	0

Note that the left cut point is offset two bits from the rule boundary, while the right cut point is offset one bit from the rule boundary. The bits within the cut points are swapped, resulting in a rule set of three rules and a rule set of one rule:

F1	F2	Class	F1	F2	Class	F1	F2	Class
100	0001	0	110	0011	**0**	**011**	**0010**	**0**
01**0**	**0100**	0						

2.2. Choosing a fitness function

In addition to selecting a good representation, it is important to define a good fitness function that rewards the right kinds of individuals. In keeping with our minimalist philosophy, we selected a fitness function involving only classification performance (ignoring, for example, length and complexity biases). The fitness of each individual rule set is computed by testing the rule set on the current set of training examples (which is typically a subset of all the examples—see section 2.6) and letting

$$fitness(individual\ i) = (percent\ correct)^2$$

This provides a bias toward correctly classifying all the examples while providing a non-linear differential reward for imperfect rule sets. This bias is equivalent to one that encourages *consistency* and *completeness* of the rule sets with respect to the training examples. A rule set is consistent when it covers no negative examples and is complete when it covers all positive examples.

2.3. GABIL: A GA-based concept learner

We are now in a position to describe GABIL, our GA-based concept learner. At the heart of this system is a GA for searching the space of rule sets for ones that perform well on a given set of positive and negative examples. Figure 1 provides a pseudo-code description of the GA used.

P(t) represents a population of rule sets. After a random initialization of the population, each rule set is evaluated with the fitness function described in section 2.2. Rule sets are probabilistically selected for survival in proportion to their fitness (i.e., how consistent and complete they are). Crossover and mutation are then applied probabilistically to the

```
procedure GA;
begin
 t = 0;
 initialize population P(t);
 fitness P(t);
 until (done)
        t = t + 1;
        select P(t) from P(t-1);
        crossover P(t);
        mutate P(t);
        fitness P(t);
end.
```

Figure 1. The GA in GABIL.

surviving rule sets, to produce a new population. This cycle continues until as consistent and complete a rule set as possible has been found within the time/space constraints given.

Traditional concept learners differ in the ways examples are presented. Many systems presume a *batch* mode, where all instances are presented to the system at once. Others work in an *incremental* mode, where one or a few of the instances are presented to the system at a time. In designing a GA-based concept learner, the simplest approach involves using a batch mode, in which a fixed set of training examples is presented and the GA must search the space of variable-length strings described above for a set of rules with high fitness (100% implies completeness and consistency on the training set).

However, in many situations learning is a never ending process in which new examples arrive incrementally as the learner explores its environment. The examples themselves can in general contain noise and are not carefully chosen by an outside agent. These are the kinds of problems that we are most interested in, and they imply that a concept learner must evolve concept descriptions incrementally from non-optimal and noisy instances.

The simplest way to produce an incremental GA concept learner is as follows. The concept learner initially accepts a single example from a pool of examples and searches for as perfect a rule set as possible for this example within the time/space constraints given. This rule set is then used to predict the classification of the next example. If the prediction is incorrect, the GA is invoked (in batch mode) to evolve a new rule set using the two examples. If the prediction is correct, the example is simply stored with the previous example and the rule set remains unchanged. As each new additional instance is accepted, a prediction is made, and the GA is rerun in batch mode if the prediction is incorrect. We refer to this mode of operation as *batch-incremental* and we refer to the GA batch-incremental concept learner as GABIL.

Although the batch-incremental mode is more costly to run than the batch, it provides a much more finely grained measure of performance that is more appropriate for situations in which learning never stops. Rather than measure an algorithm's performance using only a small training subset of the instances for learning, the batch-incremental mode measures the performance of this algorithm over *all* available instances. Therefore, every instance acts as both a testing instance and then a training instance.

Our ultimate goal is to achieve a pure incremental system that is capable of responding to even more complex situations, such as when the environment itself is changing during the learning process. In this article, however, we report on the performance of GABIL, our batch-incremental concept learner.

3. Empirical system comparisons

The experiments in this section are designed to compare the predictive performance of GABIL and four other concept learners as a function of incremental increases in the size and complexity of the target concept.

3.1. The domains

The experiments involve two domains: one artificial and one natural. Domain 1, the artificial domain, was designed to reveal trends that relate system biases to incremental increases in target concept complexity. For this domain, we invented a four-feature world in which each feature has four possible distinct values (i.e., there are 256 instances in this world).

Within Domain 1, we constructed a set of 12 target concepts. We varied the complexity of the 12 target concepts by increasing both the number of rules (disjuncts) and the number of relevant features (conjuncts) per rule required to correctly describe the concepts. The number of disjuncts ranged from one to four, while the number of conjuncts ranged from one to three. Each target concept is labeled as nDmC, where n is the number of disjuncts and m is the number of conjuncts (see appendix 2 for the definition of these target concepts).

For each of the target concepts, all 256 instances in the set were labeled as positive or negative examples of the target concept. The 256 examples were randomly shuffled and then presented sequentially in batch-incremental mode. This procedure was repeated 10 times (trials) for each concept and learning algorithm pair.

For Domain 2, we selected a natural domain to further test our conjectures about system biases. Domain 2 is a well-known natural database for diagnosing breast cancer (Michalski et al., 1986) This database has descriptions of cases for 286 patients, and each case (instance) is described in terms of nine features. There is a small amount of noise of unknown origin in the database manifested as cases with identical features but different classifications. The target concept is considerably more complex than any of the concepts in the nDmC world. For example, after seeing all 286 instances, the AQ14 system (also known as NEWGEM, described below) develops an inductive hypothesis having 25 disjuncts and an average of four conjuncts per disjunct. Since GABIL and ID5R can only handle nominals, and the breast cancer instances have features in the form of numeric intervals, we converted the breast cancer (BC) database to use nominal features. This conversion necessarily loses the inherent ordering information associated with numeric intervals. For example, the feature *age* is defined to have numeric interval values {10–19, 20–29, . . . , 90–99} in the original database, and is represented as the set {A1, A2, . . . , A9} of nominals in the converted database. When using the BC database, we again randomly shuffled the instances and averaged over 10 runs.

It should be noted that all the problems in these two test domains are single-class problems. As discussed earlier, evaluating this approach on multi-class problems is part of our future plans.

3.2. The systems

The performance of the GABIL system described in section 2.3 was evaluated on both domains. Standard GA settings of 0.6 for two-point crossover and 0.001 for mutation were used. The choice of population size was more difficult. With large, complex search spaces, larger population sizes are preferable but generally require more computation time. With our unoptimized batch-incremental version of GABIL, we were able to use a population size of 1000 for the artificial domain. However, for the BC domain, a population size of 100 was used in order to keep the experimental computation time reasonable.[4]

To better assess the GABIL system, four other concept learners were also evaluated on the target concept domains. We selected four systems to represent all four combinations of batch and incremental modes, and two popular hypothesis forms (DNF and decision trees). The chosen systems are AQ14 (Mozetic, 1985), which is based on the AQ algorithm described in Michalski (1983), C4.5 (Quinlan, unpublished), ID5R (Utgoff, 1988), and Iba's Algorithm Concept Learner (IACL) (Gordon, 1990), which is based on Iba's algorithm described in Iba (1979). AQ14 and IACL form modified DNF hypotheses. The C4.5 and ID5R systems are based on the ID algorithm described in Quinlan (1986), and form decision tree hypotheses. AQ14 and C4.5 are run in batch-incremental mode, since they are batch systems. ID5R and IACL are incremental.

AQ14, like AQ, generates classification rules from instances using a beam search. This system maintains two sets of classification rules for each concept: one set, which we call the *positive hypothesis*, is for learning the target concept, and the other set, which we call the *negative hypothesis*, is for learning the negation of the target concept. (Note that GABIL currently uses only a positive hypothesis.) AQ14, like GABIL, generates classification rules in a modified DNF that allows internal disjunction of feature values. Internal disjunction allows fewer external disjuncts in the hypotheses.

AQ14's learning method guarantees that its inductive hypotheses will be consistent and complete with respect to all training examples. The learning method, called the Star Algorithm, generates a hypothesis for each class *C*. Potential rules for this hypothesis are formed from a randomly chosen example, called a *seed*, by maximally generalizing the description of the seed without covering any examples of the wrong class. One rule is chosen from the set of potential rules, using a user-specified set of criteria, and this rule is added to the hypothesis for *C*. This procedure repeats to generate more rules for the hypothesis until all examples of class *C* are covered by the hypothesis.

AQ14's criteria for hypothesis preference (biases) influence its learning behavior. This system's performance depends on these criteria, as well as on other parameter settings. The particular parameter settings that we chose for AQ14 implement a preference for simpler inductive hypotheses, e.g., inductive hypotheses having shorter rules.[5]

C4.5 uses a decision tree representation rather than a rule representation for its inductive hypotheses. Each decision tree node is associated with an instance feature. Each node represents a test on the value of the feature. Arcs emanating from a feature node correspond to values of that feature. Each leaf node is associated with a classification (e.g., positive or negative if one concept is being learned). To view a decision tree as a positive DNF hypothesis, one would consider this hypothesis to be the disjunction of all paths (a conjunction of feature values) from the root to a positive leaf. The negative hypothesis is simply the disjunction of all paths from the root to a negative leaf.

An information-theoretic measure biases the search through the space of decision trees. Trees are constructed by first selecting a root node, then the next level of nodes, and so on. Those tree nodes, or features, that minimize entropy and therefore maximize information gain are selected first. The result is a preference for simpler (i.e., shorter) decision trees. C4.5 does not require completeness or consistency.

Two configurations of C4.5 are available: *pruned* and *unpruned*. Pruning is a process of further simplifying decision trees. This process, which occurs after the decision tree has been generated, consists of testing the tree on previously seen instances and replacing subtrees with leaves or branches whenever this replacement improves the classification accuracy.[6] Pruning is designed both for tree simplification (which increases the simplicity preference) and for improved prediction accuracy. Since it was not obvious to us when either configuration is preferable, we used both versions in our experiments.

ID5R learns with the same basic algorithm as C4.5. However, this system learns incrementally. Other than the incremental learning, ID5R's biases are nearly identical to those of C4.5 unpruned. One minor difference is that, unlike C4.5, ID5R does not predict the class of a new instance when it cannot make a prediction, e.g., when the instance is not covered by the decision tree. We have modified ID5R to make a random prediction in this case.[7]

The fourth system to be compared with GABIL is IACL (Gordon, 1990), a system similar to AQ14. IACL is not as well known as the other systems described above, and therefore we describe it in slightly more detail. IACL maintains two DNF hypotheses, one for learning the target concept, and one for learning the negation of that concept. Internal disjunction is permitted, and consistency and completeness are required. Unlike AQ14, though, IACL learns incrementally and prefers hypotheses that are more specific (i.e., less general) rather than simpler. A *merging* process maintains completeness. The merging process incorporates each new instance not already covered by a hypothesis into the hypothesis of the same class as that instance by performing a small amount of generalization. This is done by forming new hypothesis rules using a *most specific generalization* (MSG) operator. From every rule in the hypotheses, IACL forms a new rule that has feature values equal to the most specific generalization of the feature values of the new instance and those of the original rule. Each new rule is kept only if it is consistent with previous instances. Otherwise, the original rule is kept instead. If the instance cannot merge with any existing rule of the hypothesis, a description of it is added as a new rule.

When the features are nominals, as is the case for our experiments, the most specific generalization is the internal disjunction of the feature values of the rule and those of the new instance. For example, suppose the system receives its first instance, which is positive and is a small sphere. Then the initial positive hypothesis is

If ((F1 = small) and (F2 = sphere)) then it is a widget.

If the second instance is a medium cube, and it is positive, the positive hypothesis becomes

If ((F1 = small or medium) and (F2 = sphere or cube)) then it is a widget.

Note that this new hypothesis matches medium spheres and small cubes, though they have not been seen yet.

IACL's *splitting* process maintains consistency. If the system incorrectly predicts the class of a new instance, the system uses previously saved instances to relearn the hypotheses correctly. Let us continue with our example to illustrate the splitting process. Suppose the system now receives a new negative example that is a medium sphere. The current positive hypothesis matches this example, thereby violating the consistency requirement. After the splitting process, the positive hypothesis becomes

If ((F1 = small) and (F2 = sphere)) then it is a widget
or
If ((F1 = medium) and (F2 = cube)) then it is a widget,

and the negative hypothesis is

If (F1 = medium) and (F2 = sphere) then it is not a widget.

New instances can be merged with these rules to generalize the hypotheses whenever merging preserves consistency with respect to previous instances.

3.3. *Performance criteria*

We feel that learning curves are an effective means for assessing performance in the context of incremental concept learning. In the experiments reported here, each curve represents an average performance over 10 independent trials for learning a single target concept. During each trial, we keep track of a small window of recent outcomes, counting the correct predictions within that window. The value of the curve at each time step represents the percent correct achieved over the most recent window of instances. A window size of 10 was used for the artificial domain and one of size 50 for the BC domain. The sizes were chosen experimentally to balance the need for capturing recent behavior and the desire to smooth short-term variations in the learning curves.

After generating learning curves for each target concept, we collapsed the information from these curves into two performance criteria. The first, called the *prediction accuracy* (PA) criterion, is the average over all values on a learning curve, from the beginning to the end of learning a target concept. We did this to simplify the presentation of our results and to facilitate the system performance comparisons. The second performance criterion, called *convergence* (C), is the number of instances seen before a 95% prediction accuracy is maintained. If a 95% prediction accuracy cannot be achieved (e.g., on the BC database), then C is not defined. The finely grained measure obtainable with batch-incremental and incremental modes facilitates this performance criterion as well.

The criteria just described are *local* in the sense that they apply to a single target concept. For each local criterion there is a corresponding *global* criterion that considers all target concepts in a domain. The global prediction accuracy criterion is the average of the PA criteria values on every target concept within a domain. Likewise, the global convergence criterion is the average of the C criteria values on all the target concepts of a domain. Since the global criteria are based on far more data than the local criteria, we base most of our conclusions from the experiments on the former.

3.4. Results

Table 1 shows the results of the PA and global PA (denoted "Average" in the tables) criteria for measuring the performance of all systems on the nDmC and BC target concepts, while table 2 shows the results of applying the C and global C (denoted "Average" in the tables) criteria to measure performance on the nDmC concepts only (since no system achieves 95% prediction accuracy on the BC database). Although there are differences between tables 1 and 2, the general trend is similar. From these tables we can see that AQ14 is the best performer overall. In particular, AQ14 is the top or close to the top performer on the nDmC concepts. This system does not, however, perform as well as the other systems on the BC target concept (except IACL). These results are due to the fact that AQ14, when using our chosen parameter settings, is a system that is well tuned to simpler DNF target concepts.[8]

IACL does not perform as well as the other systems on the BC target concept. We consider this to be a result of IACL's sensitivity to our conversion of numeric intervals to nominals, as was discussed earlier. IACL's MSG operator is well suited for learning when the instance features are structured nominals (i.e., have generalization trees to structure their values) or numeric, but is not well suited for learning when the features are (unstructured) nominals. According to Gordon (1990), IACL performs very well on the numeric form of the BC database.[9] Other experiments, which are not reported here, indicate that the

Table 1. Prediction accuracy.

	Prediction Accuracy					
TC	AQ14	C4.5P	C4.5U	ID5R	IACL	GABIL
1D1C	99.8	98.5	99.8	99.8	98.1	95.2
1D2C	98.4	96.1	99.1	99.0	96.7	95.8
1D3C	97.4	98.5	99.0	99.1	90.4	95.7
2D1C	98.6	93.4	98.2	97.9	95.6	92.0
2D2C	96.8	94.3	98.4	98.2	94.5	92.7
2D3C	96.7	96.9	97.6	97.9	95.3	94.6
3D1C	98.0	78.8	92.4	91.2	93.2	90.4
3D2C	95.5	92.2	97.4	96.7	92.1	90.3
3D3C	95.3	95.4	96.6	95.6	94.9	92.8
4D1C	95.8	66.4	77.0	70.2	92.3	89.6
4D2C	93.8	90.5	95.2	81.3	89.5	87.4
4D3C	93.5	93.8	95.1	90.3	94.2	88.9
Average	96.6	91.2	95.5	93.1	93.9	92.1
BC	60.5	72.4	65.9	63.4	60.1	68.7

Table 2. Convergence to 95%.

	Convergence					
TC	AQ14	C4.5P	C4.5U	ID5R	IACL	GABIL
1D1C	13	37	14	12	33	87
1D2C	28	155	24	26	91	100
1D3C	57	0	0	0	96	96
2D1C	28	100	37	44	61	109
2D2C	43	126	32	40	139	148
2D3C	86	181	86	75	134	249
3D1C	34	253	149	137	203	103
3D2C	78	122	45	52	141	125
3D3C	195	253	135	123	125	225
4D1C	82	253	253	255	222	131
4D2C	78	113	55	255	188	142
4D3C	154	253	134	224	138	229
Average	73	154	80	104	131	145

conversion of the BC data to a nominal form does not adversely affect performance for AQ14 and C4.5.

C4.5 pruned (abbreviated C4.5P in the tables) performs well on all but the target concepts that have many short disjuncts. On 4D1C, which has the most short disjuncts of any target concept in the artificial domain, all ID-based systems (C4.5 pruned and unpruned, as well as ID5R) perform poorly.[10] Based on the global performance criteria, C4.5 unpruned (abbreviated C4.5U in the tables) performs the best of the ID-based systems on the artificial domain, whereas C4.5 pruned performs the best on the BC domain.

GABIL appears to be a good overall performer. It does not do superbly on any particular concept, but it also does not have a distinct region of the space of concepts on which it clearly degrades. Furthermore, GABIL is quite competitive on the difficult BC target concept. The statistical significance of these results is presented in appendix 1.

It is clear from these results that none of the systems under evaluation is superior to all others on all the target concepts. The dominance of one technique on a certain class of concepts appears to be due in large part to the built-in forms of bias it embodies, and these can have a negative impact on other classes of concepts.

4. A more robust concept learner

The GABIL system evaluated above incorporates a "pure" GA as its search mechanism in the sense that there were no specific changes made to the internal representation or genetic operators relating to the task of concept learning. As in other application tasks, this generally results in a good overall GA-based problem solver, but one that can be out-performed by task-specific approaches, particularly on simpler problems (see, for example, De Jong & Spears, 1989). However, one of the nice properties of a GA-based system is that it is not difficult to augment GAs with task-specific features to improve performance on that task.

```
procedure GA;
begin
 t = 0;
 initialize population P(t);
 fitness P(t);
 until (done)
        t = t + 1;
        select P(t) from P(t-1);
        crossover P(t);
        mutate P(t);
        new_op1 P(t);    /* additional operators */
        new_op2 P(t);
        ...
        fitness P(t);
end.
```

Figure 2. Extending GABIL's GA operators.

After obtaining the performance comparisons described in the previous section, we felt that extending GABIL with a small set of features appropriate to concept learning would significantly enhance its overall performance and robustness. Our approach was as follows. Each of the traditional concept learners evaluated above appeared to contain one or more biases that we considered to be largely responsible for that system's success on a particular class of target concepts. We selected a subset of these biases to be implemented as additional "genetic" operators to be used by GABIL's GA search procedure (see figure 2). The virtue of this approach is the simple and unified way GABIL's underlying search process can be extended to include various traditional forms of concept learning bias.

Since AQ14 seemed to be the best overall performer, we selected it as our initial source of additional operators for GABIL. As we have described above, the AQ14 system used in our experiments has preferences for simpler and more general rules. After studying the results of our initial evaluation, we hypothesized that this is one of the biases largely responsible for AQ14's superior performance on the *n*D*m*C concepts. This analysis led to the addition of two new GABIL operators that add biases for simpler and more general descriptions.

4.1. The adding alternative operator

One of the mechanisms AQ uses to increase the generality of its inductive hypotheses is the "adding alternative" generalization operator of Michalski (1983). This operator generalizes by adding a disjunct (i.e., alternative) to the current classification rule. The most useful form of this operator, according to Michalski (1983), is the addition of an internal disjunct. For example, if the disjunct is

(F1 = small) and (F2 = sphere)

then the adding alternative operator might create the new disjunct

(F1 = small) and (F2 = sphere or cube).

This operator, which we call AA (the *adding alternative operator*), is easily implemented in GABIL by including an additional mutation that, unlike the normal mutation operator, has an asymmetric mutation rate. In particular, in the studies reported here, this operator incorporates a 75% probability of mutating a bit to a 1, but a 25% probability of mutating it to a 0. Therefore, the AA operator in GABIL has a strong preference for adding internal disjuncts. To illustrate, the adding alternative operator might change the disjunct

F1	F2
100	**100**

to

F1	F2
100	**110**

Note that feature F2 has been generalized in this disjunct.

As with the other genetic operators, the adding alternative operator is applied probabilistically to a subset of the population each generation. In the studies reported here it was applied at a rate of 0.01 (1%).[11] For clarity in reporting the experimental results, we call the version of GABIL with the adding alternative operator "GABIL+A."

4.2. The dropping condition operator

A second, and complementary, generalization mechanism leading to simpler hypotheses involves removing what appear to be nearly irrelevant conditions from a disjunct. This operator, which we call DC (the *dropping condition operator*), is based on the generalization operator of the same name described in Michalski (1983). For example, if the disjunct is

(F1 = small or medium) and (F2 = sphere)

then the DC operator might create the new disjunct

(F2 = sphere).

The DC operator is easily added to GABIL in the following manner. When this operator is applied to a particular member of the population (i.e., a particular rule set), each disjunct is deterministically checked for possible condition dropping. The decision to drop a condition is based on a criterion from Gordon (1990) and involves examining the bits of each feature in each disjunct. If more than half of the bits of a feature in a disjunct are 1's, then the remaining 0 bits are changed to 1's. By changing the feature to have all 1 values, this operator forces the feature to become irrelevant within that disjunct and thereby simulates the effect of a shortened disjunct. To illustrate, suppose this operator is applied to the following disjunct:

F1	F2
110	100

```
procedure GA;
begin
 t = 0;
 initialize population P(t);
 fitness P(t);
 until (done)
        t = t + 1;
        select P(t) from P(t-1);
        crossover P(t);
        mutate P(t);
        add_altern P(t);    /* +A */
        drop_cond P(t);     /* +D */
        fitness P(t);
end.
```

Figure 3. Extended GABIL.

Then the dropping condition operator will result in a new disjunct as follows:

F1	F2
111	100

Note that feature Fl is now irrelevant within this disjunct.

As with the other genetic operators, this new operator is applied probabilistically to a subset of the population each generation. In the experiments reported here, a rate of 0.60 (60%) was used. We make no claim that the rates used for either of these new operators are in any sense optimal. In these studies we selected a rate that seemed reasonable on the basis of a few informal experiments. Our preference (see section 5) is that such things be self-adaptive.

We call GABIL with the DC operator "GABIL+D." When both task-specific operators are added to GABIL, the resulting system is called "GABIL+AD" (see figure 3). Note that there is an interesting complementary relationship between these two operators in that adding alternatives can set the stage for dropping a condition altogether.

The augmented forms of GABIL do not change in any way the overall structure of the GA-based system described earlier (compare figures 1 and 3). The only difference is that the set of "genetic" operators has been expanded. The result is that, after the traditional crossover and mutation operators have been used in the normal manner to produce new offspring (rule sets) from parents, the two new operators are (probabilistically) applied to each offspring, producing additional task-specific changes.

4.3. Results

To study the effects of adding these bias operators to GABIL, GABIL+A, GABIL+D, and GABIL+AD have been run on the same concept domains used earlier. Table 3 shows the results of system performance measured using the PA and global PA criteria. Table 4 shows the results of system performance measured using the C and global C criteria. GABIL is abbreviated "G" in the tables.

Table 3. Prediction accuracy.

	Prediction Accuracy			
TC	GABIL	G+A	G+D	G+AD
1D1C	95.2	96.1	97.7	97.7
1D2C	95.8	96.2	97.4	97.3
1D3C	95.7	95.7	96.7	96.7
2D1C	92.0	93.1	97.4	97.0
2D2C	92.7	95.0	96.3	96.9
2D3C	94.6	94.5	95.8	95.0
3D1C	90.4	91.9	96.0	96.6
3D2C	90.3	91.6	94.5	94.6
3D3C	92.8	92.7	94.2	92.9
4D1C	89.6	90.9	95.1	95.2
4D2C	87.4	89.7	93.0	92.7
4D3C	88.9	89.2	92.3	90.0
Average	92.1	93.1	95.5	95.2
BC	68.7	69.1	71.5	72.0

Table 4. Convergence to 95%.

	Convergence			
TC	GABIL	G+A	G+D	G+AD
1D1C	87	58	28	32
1D2C	100	85	59	68
1D3C	96	97	94	97
2D1C	109	90	42	42
2D2C	148	93	82	55
2D3C	249	250	136	250
3D1C	103	104	54	39
3D2C	125	127	76	62
3D3C	225	240	161	240
4D1C	131	120	67	62
4D2C	142	133	75	75
4D3C	229	253	166	248
Average	145	138	87	106

According to the global criteria in tables 3 and 4, GABIL+A does not perform as well as GABIL+D or GABIL+AD. On the BC target concept, the combination of both operators (GABIL+AD) is the best. It is interesting to note, however, that on the *n*D*m*C domain, GABIL+AD does not perform as well as GABIL+D.

These results indicate that one can improve GABIL's performance on certain classes of concepts by the addition of an appropriate set of bias operators. On the other hand, it is not possible in general to know beforehand which set of biases is best. These results also point out the danger of indiscriminately including multiple biases as a strategy for overcoming this lack of a priori knowledge, since multiple simultaneous biases can in fact interfere with one another, leading to a degradation in performance. These results, which confirm similar bias problems exhibited in other contexts, motivated us to focus on a more sophisticated way of improving GABIL's overall robustness, namely, by having it dynamically adjust its own task-specific biases.

5. An adaptive GA concept learner

Although genetic algorithms themselves represent a robust adaptive search mechanism, most GA implementations involve static settings for such things as the population size, the kind of representation and operators used, and the operator probabilities. There have been a number of attempts to make these aspects of GAs more adaptive. We provide a brief overview of this work in the next section.

5.1. Adaptive GAs

There have been two approaches to building more adaptive GAs, which we refer to as the *within-problem* approach and the *across-problem* approach. The within-problem approach adapts a GA dynamically, as it solves one problem. In contrast, the across-problem approach adapts GAs over the course of many problems. One good example of the across-problem approach is provided by Grefenstette (1986). In that paper, a separate *meta*-GA is used to adapt a GA as it solves a suite of problems. The advantage of such an approach is that the resulting system performs robustly on a suite of problems. Unfortunately, the approach is also time consuming, since each problem must be solved a large number of times. Furthermore, the adaptation is coarse, in the sense that the system is not necessarily optimal on any given problem. Within-problem adaptation provides a finer-grained approach, since the GA is adapted while one problem is solved. Furthermore, since the problem is solved only once, the approach can require much less time. We concentrate on the within-problem approach, since we wish to adapt the GA as it solves each concept learning problem.

Within-problem approaches can be further divided into two categories, *coupled* and *uncoupled*, based on the observation that an adaptive GA is in effect searching two spaces: the original problem space, and the space of adaptations to the underlying GA itself. The relationship of these two search processes is an important design consideration for adaptive GAs.

In a coupled approach, both searches are handled simultaneously by a single GA search procedure. This is accomplished by using the underlying population to store information relevant to the adaptive mechanism as well as the standard information regarding the original problem space being searched. This approach is elegant and straightforward, since no new adaptive mechanism is required (see Schaffer et al. (1987) for examples of this approach). Unfortunately, this coupling also means that the additional search can be hindered by the same issues that hinder the search of the problem space. For example, one possible concern is that this mechanism may only work well with large population sizes. As with any other statistical sampling algorithm, small populations (samples) may be misleading and lead to wrong conclusions. This issue will be raised again later in this article.

An uncoupled approach does not rely upon the GA for the adaptive mechanism. Rather, the behavior of the underlying GA is adjusted by a separate control mechanism (see Davis

(1989) and Janikow (1991) for examples). While this may alleviate the problems associated with coupling, such mechanisms appear to be difficult to construct, and involve complicated bookkeeping. Although we may explore this route in future work, we concentrate on the conceptually simpler coupled approach in this article. We next consider how to implement a coupled within-problem approach within GABIL.

5.2. Adaptive GABIL

Recall that the task-specific operators added to GABIL (DC and AA) were added in a static way. That is, they were either present or not present for an entire experiment. If they were present, they were applied via fixed probabilities to all new individuals. The simplest coupled way to make the selection and application of these operators more adaptive is to have each individual specify which operators can be applied to it. The intuition here is that those individuals that enable the "correct" operators will be more fit from a survival point of view. The result should be a system capable of performing the search for the best set of operators (biases) and the search for the best hypotheses in parallel (see Baeck et al. (1991) for related work).

Such an approach is easily implemented by adding to each individual additional control bits (one for each adaptive operator). Each bit determines whether the corresponding operator can be used on that individual. If the control bit is 0, the associated operator is not permissible, and cannot be fired (thus ignoring the operator probability). If the control bit is 1, the associated operator is permissible, and fires according to the relevant operator probability. These control bits act as added Boolean preconditions for the operators. The values of the control bits are evolved in the normal way through selection, crossover, and mutation.[12]

As an initial test of this approach, GABIL was modified to include two extra control bits, one for each of the task-specific operators introduced earlier. For example, consider the following rule set:

F1	F2	Class	F1	F2	Class	D	A
010	001	0	110	011	0	1	0

The two added control bits are indicated with the letters "D" and "A" (for dropping condition and adding alternative, respectively). For this rule set the dropping condition operator is permissible, while the adding alternative operator is not. So, for example, the DC operator would change the rule set to

F1	F2	Class	F1	F2	Class	D	A
010	001	0	111	111	0	1	0

We call this modified system "adaptive GABIL," and have begun to explore its potential for effective dynamic bias adjustment. To get an immediate and direct comparison with the earlier results, adaptive GABIL was run on the *n*D*m*C and BC target concepts. The results are presented in tables 5 and 6.

Table 5. Prediction accuracy.

	Prediction Accuracy				
TC	GABIL	G+A	G+D	G+AD	Adaptive
1D1C	95.2	96.1	97.7	97.7	97.6
1D2C	95.8	96.2	97.4	97.3	97.4
1D3C	95.7	95.7	96.7	96.7	96.5
2D1C	92.0	93.1	97.4	97.0	96.1
2D2C	92.7	95.0	96.3	96.9	96.2
2D3C	94.6	94.5	95.8	95.0	95.4
3D1C	90.4	91.9	96.0	96.6	95.9
3D2C	90.3	91.6	94.5	94.6	94.0
3D3C	92.8	92.7	94.2	92.9	94.7
4D1C	89.6	90.9	95.1	95.2	95.8
4D2C	87.4	89.7	93.0	92.7	92.8
4D3C	88.9	89.2	92.3	90.0	92.1
Average	92.1	93.1	95.5	95.2	95.4
BC	68.7	69.1	71.5	72.0	70.3

Table 6. Convergence to 95%.

	Convergence				
TC	GABIL	G+A	G+D	G+AD	Adaptive
1D1C	87	58	28	32	34
1D2C	100	85	59	68	58
1D3C	96	97	94	97	97
2D1C	109	90	42	42	50
2D2C	148	93	82	55	80
2D3C	249	250	136	250	120
3D1C	103	104	54	39	53
3D2C	125	127	76	62	70
3D3C	225	240	161	240	128
4D1C	131	120	67	62	55
4D2C	142	133	75	75	80
4D3C	229	253	166	248	130
Average	145	138	87	106	80

The results of the global criteria, shown at the bottom of tables 5 and 6, highlight a couple of important points. First, on the nDmC domain, the adaptive GABIL outperforms the original GABIL, GABIL+A, and GABIL+AD. Furthermore, the adaptive GABIL performs almost as well as GABIL+D from a prediction accuracy criterion, and better from a convergence criterion. Adaptive GABIL outperforms GABIL+AD, particularly from the standpoint of the global C criterion. This shows the danger of indiscriminately including multiple fixed biases, which can interfere with each other, producing lower performance. These results demonstrate the virtues of adaptive GABIL in selecting the appropriate biases.

On the BC target concept, adaptive GABIL performs better than GABIL and GABIL+A, but is worse than GABIL+D and GABIL+AD. This suggests that adaptive GABIL's advantage is diminished when smaller population sizes (e.g., population sizes of 100) are involved. To address this issue, future versions of GABIL will have to adapt the population size, as well as operator selection.

In comparison to the other systems, the new adaptive GABIL is much better than C4.5 on the *n*D*m*C domain, and close on the BC target concept. Also, adaptive GABIL is competitive with AQ14 on the *n*D*m*C domain and is much better on the BC target concept. We have tested the statistical significance of these results (see appendix 1) and have found that when adaptive GABIL outperforms other systems, the results are generally significant (at a 90% level). Furthermore, when other systems outperform adaptive GABIL, the results are generally not significant (i.e., significance is 80% or lower). The only two notable exceptions are on the BC database. Both C4.5 and GABIL+AD outperform adaptive GABIL at a 95% level of significance. We believe that the latter exception is due to the small population size (100). The former exception will be addressed when we incorporate C4.5's information-theoretic biases into GABIL. This bias can be quite easily implemented as a "genetic" operator by making features with higher entropy values more likely to have 1's (since higher entropy values imply less relevance).

An interesting question at this point is whether the improved performance of adaptive GABIL is the result of any significant bias adjustment during a run. This is easily monitored and displayed. Figures 4 and 5 illustrate the frequency with which the dropping condition (DC) and adding alternative (AA) operators are used by adaptive GABIL for two target

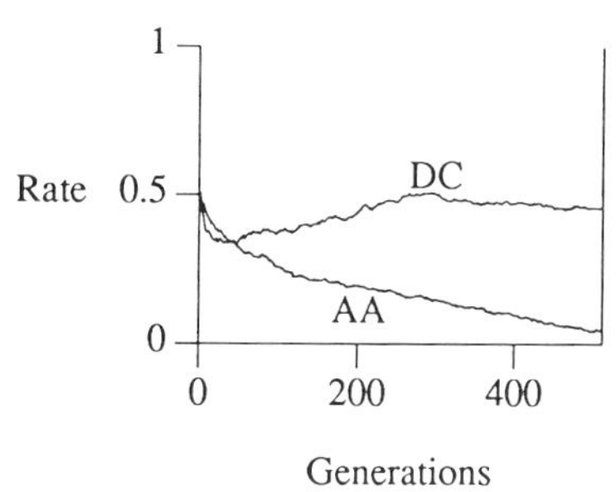

Figure 4. 3D3C.

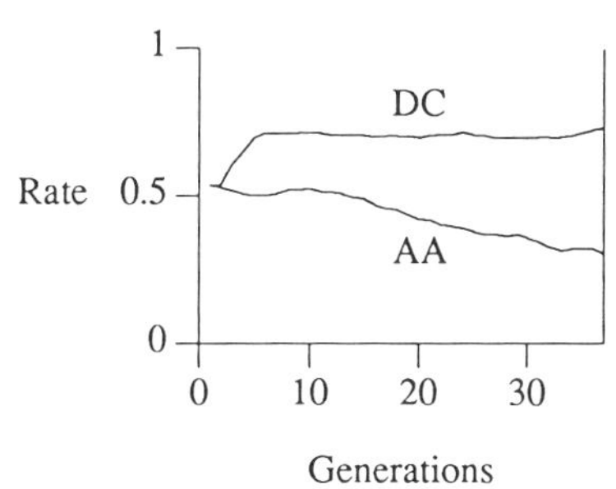

Figure 5. 4D1C.

concepts: 3D3C and 4D1C. Since the control bits for each operator are randomly initialized, roughly half of the initial population contain positive control bits resulting in both operators starting out at a rate of approximately 0.5. As the search progresses towards a consistent and complete hypothesis, however, these frequencies are adaptively modified. For both target concepts, the DC operator is found to be the most useful, and consequently is fired with a higher frequency. This is consistent with table 5, which indicates that GABIL+D outperforms GABIL+A. Furthermore, note the difference in predictive accuracy between GABIL and GABIL+D on the two target concepts. The difference is greater for the 4D1C target concept, indicating the greater importance of the DC operator. This is reflected in figures 4 and 5, in which the DC operator evolves to a higher firing frequency on the 4D1C concept, in comparison with the 3D3C concept. Similar comparisons can be made with the AA operator.

Considering that GABIL is now clearly performing the additional task of selecting appropriate biases, these results are very encourging. We are in the process of extending and refining GABIL as a result of the experiments described here. We are also extending our experimental analysis to include other systems that attempt to dynamically adjust their bias.

6. Related work on bias adjustment

Adaptive bias, in our context, is similar to dynamic preference (bias) adjustment for concept learning (see Gordon (1990) for related literature). The vast majority of concept learning systems that adjust their bias focus on changing their representational bias. The few notable exceptions that adjust the algorithmic bias include the Competitive Relation Learner and Induce and Select Optimizer combination (CRL/ISO) (Tcheng et al., 1989), Climbing in the Bias Space (ClimBS) (Provost, 1991), PEAK (Holder, 1990), the Variable Bias Management System (VBMS) (Rendell et al., 1987), and the Genetic-based Inductive Learner (GIL) (Janikow, 1991).

We can classify these systems according to the type of bias that they select. Adaptive GABIL shifts its bias by dynamically selecting generalization operators. The set of biases considered by CRL/ISO includes the strategy for predicting the class of new instances and the method and criteria for hypothesis selection. The set of biases considered by ClimBS includes the beam width of the heuristic search, the percentage of the positive examples a satisfactory rule must cover, the maximum percentage of the negative examples a satisfactory rule may cover, and the rule complexity. PEAK's changeable algorithmic biases are learning algorithms. They are rote learning, empirical learning (with a decision tree), and explanation-based generalization (EBG). GIL is most similar to GABIL, since it also selects between generalization operators. However, it does not use a GA for that selection and only uses a GA for the concept learning task.

We can also classify these systems according to whether or not their searches through the space of hypotheses and the space of biases are coupled. GABIL is unique along this dimension because it is the only system that couples these searches. The advantages and disadvantages of a coupled approach were presented in section 5. We summarize these comparisons in table 7.

Table 7. Comparison of system characteristics.

System	Bias Space	Searches	GA
CRL/ISO	hypothesis selection criteria, prediction strategy	uncoupled	no
ClimBS	beam width, hypothesis coverage	uncoupled	no
PEAK	learning strategy	uncoupled	no
VBMS	concept learners	uncoupled	no
GIL	generalization operators	uncoupled	yes
Adaptive GABIL	generalization operators	coupled	yes

The VBMS system is different from the others mentioned above. The primary task of this system is to identify the concept learner (which implements a particular set of algorithmic biases) that is best suited for each problem along a set of problem-characteristic dimensions. Problem-characteristic dimensions that this system considers are the number of training instances and the number of features per instance. Three concept learners are tested for their suitability along these problem-characteristic dimensions. VBMS would be an ideal companion to any of the above-mentioned systems. This system could map out the suitability of biases to problems, and then this knowledge could be passed on to the other systems to use in an initialization procedure for constraining their bias space search.

7. Discussion and future work

We have presented a method for using genetic algorithms as a key element in designing robust concept learning systems and have used this approach to implement a system that compares favorably with other concept learning systems on a variety of target concepts. We have shown that, in addition to providing a minimally biased yet powerful search strategy, the GABIL architecture allows for adding task-specific biases in a very natural way in the form of additional "genetic" operators, resulting in performance improvements on certain classes of concepts. However, the experiments in this article highlight that no one fixed set of biases is appropriate for all target concepts. In response to these observations, we have shown that this approach can be further extended to produce a concept learner that is capable of dynamically adjusting its own bias in response to the characteristics of the particular problem at hand. Our results indicate that this is a promising approach for building concept learners that do not require a "human in the loop" to adapt and adjust the system to the requirements of a particular class of concepts.

The current version of GABIL adaptively selects between two forms of bias taken from a single system (AQ14). In the future, we plan to extend this set of biases to include additional biases from AQ14 and other systems. For example, we would like to implement in GABIL an information-theoretic bias, which we believe is primarily responsible for C4.5's successes.

The results presented here have all involved single-class learning problems. An important next step is to extend this method to multi-class problems. We have also been focusing on adjusting the lower-level biases of learning systems. We believe that these same techniques can also be applied to the selection of higher-level mechanisms such as induction and analogy. Our final goal is to produce a robust learner that dynamically adapts to changing concepts and noisy learning conditions, both of which are frequently encountered in realistic environments.

Acknowledgments

We would like to thank the Machine Learning Group at NRL for their useful comments about GABIL, J.R. Quinlan for C4.5, and Zianping Zhang and Ryszard Michalski for AQ14.

Notes

1. Excellent introductions to GAs can be found in Holland (1975) and Goldberg (1989).
2. Greene & Smith (1987) and Janikow (1991) have also used the Pittsburgh approach. See Wilson (1987) and Booker (1989) for examples of the Michigan approach.
3. We are also investigating the use of a uniform crossover operator that has been recently shown to be more effective in certain contexts than two-point crossover.
4. Our unoptimized batch-incremental version of GABIL is somewhat slower than C4.5, AQ, and IACL. It is substantially slower than ID5R. One should not conclude from this, however, that GA concept learners are inherently slower. See Janikow (1991) for details.
5. The precise criteria used are as follows: the positive and negative inductive hypotheses are allowed to intersect provided the intersection covers no instances, noisy examples are considered positive, the maximum beam width is set to 20, and the minimum number of features and values are preferred in each rule. Other settings, which have less impact on system performance, are left at default values.
6. The type of pruning in C4.5 is a variant of *pessimistic pruning* described by Quinlan (1987) that prunes a tree to either a subtree or a leaf node (Quinlan, personal communication).
7. ID5R, like GABIL, is a research tool and therefore does not handle some of the realistic data characteristics (e.g., missing feature values) that can be handled by sophisticated systems such as C4.5.
8. AQ14 does not use flexible (partial) matching of hypotheses of instances. Flexible matching tends to improve the performance of the AQ systems (Michalski, 1990). The newest version of AQ (AQTT-15), which uses flexible matching, was unavailable at the time of this study. In the future, we plan to run AQTT-15 on our suite of target concepts.
9. When run in batch mode on the numeric BC database, with 70% of the instances in the training set and 30% in the test set, 72% of the predictions made on the test set were correct predictions (see Gordon, 1990).
10. An explanation of the difficulty of systems based on ID3 on target concepts of this type is in De Jong and Spears (1991).
11. Note that this is in addition to the standard mutation operator, which continues to fire with a probability of .001.
12. The dropping condition and adding alternative operators do not alter these control bits.

Appendix 1: Statistical significance

The following three tables give statistical significance results. Table 8 compares adaptive GABIL with all other systems on the *n*D*m*C domain (using predictive accuracy). Table 9 makes the same comparison with the convergence criterion. Table 10 compares adaptive GABIL with all other systems on the BC target concept. The column *Sig* denotes the level of significance of each comparison. The *Wins* column is "Yes" if adaptive GABIL outperformed the other system; otherwise it is "No."

In comparison with all other systems, adaptive GABIL has 19 wins, 7 losses, and 1 tie. At the 90% level of significance, 11 wins and 2 losses are significant. In comparison with the non-GA systems, adaptive GABIL has 10 wins, 4 losses, and 1 tie. Again, at the 90% level of significance, 6 wins and 1 loss are significant.

Table 8. Predictive accuracy on *n*D*m*C.

System	Sig	Wins
AQ14	80%	No
C4.5P	80%	Yes
C4.5U	<80%	No
ID5R	<80%	Yes
IACL	80%	Yes
GABIL	95%	Yes
G+A	95%	Yes
G+D	<80%	No
G+AD	<80%	Yes

Table 9. Convergence on *n*D*m*C.

System	Sig	Wins
AQ14	<80%	No
C4.5P	95%	Yes
C4.5U	<80%	Tie
ID5R	<80%	Yes
IACL	95%	Yes
GABIL	95%	Yes
G+A	95%	Yes
G+D	<80%	Yes
G+AD	<80%	Yes

Table 10. Predictive accuracy on BC.

System	Sig	Wins
AQ14	95%	Yes
C4.5P	95%	No
C4.5U	95%	Yes
ID5R	95%	Yes
IACL	95%	Yes
GABIL	90%	Yes
G+A	80%	Yes
G+D	80%	No
G+AD	95%	No

Appendix 2: Artificial domain target concepts

This appendix fully describes the target concepts of the artificial domain. There are four features, denoted as F1, F2, F3, and F4. Each feature has four values {v1, v2, v3, v4}.

All the target concepts have the following form:

4D*m*C == d1 v d2 v d3 v d4
3D*m*C == d1 v d2 v d3
2D*m*C == d1 v d2
1D*m*C == d1

For the *n*D3C target concepts, we have

d1 == (F1 = v1) & (F2 = v1) & (F3 = v1)
d2 == (F1 = v2) & (F2 = v2) & (F3 = v2)
d3 == (F1 = v3) & (F2 = v3) & (F3 = v3)
d4 == (F1 = v4) & (F2 = v4) & (F3 = v4)

For the *n*D2C target concept, we have

d1 == (F1 = v1) & (F2 = v1)
d2 == (F1 = v2) & (F2 = v2)
d3 == (F1 = v3) & (F2 = v3)
d4 == (F1 = v4) & (F2 = v4)

Finally, we define the *n*D1C target concepts:

d1 == (F1 = v1)
d2 == (F1 = v2)
d3 == (F1 = v3)
d4 == (F1 = v4)

References

Baeck, T., Hoffmeister, F., & Schwefel, H. (1991). A survey of evolution strategies. *Proceedings of the Fourth International Conference on Genetic Algorithms* (pp. 2–9). La Jolla, CA: Morgan Kaufmann.

Booker, L. (1989). Triggered rule discovery in classifier systems. *Proceedings of the Third International Conference on Genetic Algorithms* (pp. 265–274). Fairfax, VA: Morgan Kaufmann.

Davis, L. (1989). Adapting operator probabilities in genetic algorithms. *Proceedings of the Third International Conference on Genetic Algorithms* (pp. 61–69). Fairfax, VA: Morgan Kaufmann.

De Jong, K. (1987). Using genetic algorithms to search program spaces. *Proceedings of the Second International Conference on Genetic Algorithms* (pp. 210–216). Cambridge, MA: Lawrence Erlbaum.

De Jong, K., & Spears, W. (1989). Using genetic algorithms to solve NP-complete problems. *Proceedings of the Third International Conference on Genetic Algorithms* (pp. 124–132). Fairfax, VA: Morgan Kaufmann.

De Jong, K., & Spears, W. (1991). Learning concept classification rules using genetic algorithms. *Proceedings of the Twelfth International Joint Conference on Artificial Intelligence* (pp. 651–656). Sydney, Australia: Morgan Kaufmann.

Goldberg, D. (1989). *Genetic algorithms in search, optimization, and machine learning*. New York: Addison-Wesley.

Gordon, D. (1990). *Active bias adjustment for incremental, supervised concept learning*. Doctoral dissertation, Computer Science Department, University of Maryland, College Park, MD.

Greene, D., & Smith, S. (1987). A genetic system for learning models of consumer choice. *Proceedings of the Second International Conference on Genetic Algorithms* (pp. 217–223). Cambridge, MA: Lawrence Erlbaum.

Grefenstette, John J. (1986). Optimization of control parameters for genetic algorithms. *IEEE Transactions on Systems, Man, and Cybernetics, SMC-16(1)*, 122–128.
Grefenstette, John J. (1989). A system for learning control strategies with genetic algorithms. *Proceedings of the Third International Conference on Genetic Algorithms* (pp. 183–190). Fairfax, VA: Morgan Kaufmann.
Holder, L. (1990). The general utility problem in machine learning. *Proceedings of the Seventh International Conference on Machine Learning* (pp. 402–410). Austin, TX: Morgan Kaufmann.
Holland, J. (1975). *Adaptation in natural and artificial systems*. Ann Arbor, MI: The University of Michigan Press.
Holland, J. (1986). Escaping brittleness: The possibilities of general-purpose learning algorithms applied to parallel rule-based systems. In R. Michalski, J. Carbonell, & T. Mitchell (Eds.), *Machine learning: An artificial intelligence approach*. Los Altos, CA: Morgan Kaufmann.
Iba, G. (1979). *Learning disjunctive concepts from examples* (A.I. Memo 548). Cambridge, MA: Massachusetts Institute of Technology.
Janikow, C. (1991). *Inductive learning of decision rules from attribute-based examples: A knowledge-intensive genetic algorithm approach* (TR91-030). Chapel Hill, NC: The University of North Carolina at Chapel Hill, Department of Computer Science.
Koza, J.R. (1991). Concept formation and decision tree induction using the genetic programming paradigm. In H.P. Schwefel & R. Maenner (Eds.), *Parallel problem solving from nature*. Berlin: Springer-Verlag.
Michalski, R. (1983). A theory and methodology of inductive learning. In R. Michalski, J. Carbonell, & T. Mitchell (Eds.), *Machine learning: An artificial intelligence approach*. Palo Alto: Tioga.
Michalski, R. (1990). Learning flexible concepts: Fundamental ideas and a method based on two-tiered representation. In Y. Kodratoff & R. Michalski (Eds.), *Machine learning: An artificial intelligence approach*. San Mateo, CA: Morgan Kaufmann.
Michalski, R., Mozetic, L., Hong, J., & Lavrac, N. (1986). The AQ15 inductive learning system: An overview and experiments (Technical Report Number UIUCDCS-R-86-1260). Urbana-Champaign, IL: University of Illinois.
Mozetic, I. (1985). NEWGEM: Program for learning from examples, program documentations and user's guide (Report Number UIUCDCS-F-85-949). Urbana-Champaign, IL: University of Illinois.
Provost, F. (1991). *Navigation of an extended bias space for inductive learning*. Ph.D. thesis proposal, Computer Science Department, University of Pittsburgh, Pittsburgh, PA.
Quinlan, J. (1986). Induction of decision trees. *Machine Learning, 1(1)*, 81–106.
Quinlan, J. (1989). Documentation and user's guide for C4.5. (unpublished).
Rendell, L. (1985). Genetic plans and the probabilistic learning system: Synthesis and results. *Proceedings of the First International Conference on Genetic Algorithms* (pp. 60–73). Pittsburgh, PA: Lawrence Erlbaum.
Rendell, L., Seshu, R., & Tcheng, D. (1987). More robust concept learning using dynamically-variable bias. *Proceedings of the Fourth International Workshop on Machine Learning* (pp. 66–78). Irvine, CA: Morgan Kaufmann.
Schaffer, J. David, & Morishima, A. (1987). An adaptive crossover distribution mechanism for genetic algorithms. *Proceedings of the Second International Conference on Genetic Algorithms* (pp. 36–40). Cambridge, MA: Lawrence Erlbaum.
Smith, S. (1983). Flexible learning of problem solving heuristics through adaptive search. *Proceedings of the Eighth International Joint Conference on Artificial Intelligence* (pp. 422–425). Karlsruhe, Germany: William Kaufmann.
Tcheng, D., Lambert, B., Lu, S., & Rendell, R. (1989). Building robust learning systems by combining induction and optimization. *Proceedings of the Eleventh International Joint Conference on Artificial Intelligence* (pp. 806–812). Detroit, MI: Morgan Kaufmann.
Wilson, S. (1987). Quasi-Darwinian learning in a classifier system. *Proceedings of the Fourth International Workshop on Machine Learning* (pp. 59–65). Irvine, CA: Morgan Kaufmann.
Utgoff, P. (1988). ID5R: An incremental ID3. *Proceedings of the Fifth International Conference on Machine Learning* (pp. 107–120). Ann Arbor, MI: Morgan Kaufmann.

Received November 19, 1991
Accepted April 13, 1992
Final Manuscript July 23, 1992

Machine Learning, 13, 189-228 (1993)

A Knowledge-Intensive Genetic Algorithm for Supervised Learning

CEZARY Z. JANIKOW JANIKOW@RADOM.UMSL.EDU
UMSL, Department of Mathematics and Computer Science, St. Louis, MO 63121

Abstract. Supervised learning in attribute-based spaces is one of the most popular machine learning problems studied and, consequently, has attracted considerable attention of the genetic algorithm community. The full-memory approach developed here uses the same high-level descriptive language that is used in rule-based systems. This allows for an easy utilization of inference rules of the well-known inductive learning methodology, which replace the traditional domain-independent operators and make the search task-specific. Moreover, a closer relationship between the underlying task and the processing mechanisms provides a setting for an application of more powerful task-specific heuristics. Initial results obtained with a prototype implementation for the simplest case of single concepts indicate that genetic algorithms can be effectively used to process high-level concepts and incorporate task-specific knowledge. The method of abstracting the genetic algorithm to the problem level, described here for the supervised inductive learning, can be also extended to other domains and tasks, since it provides a framework for combining recently popular genetic algorithm methods with traditional problem-solving methodologies. Moreover, in this particular case, it provides a very powerful tool enabling study of the widely accepted but not so well understood inductive learning methodology.

Keywords. Genetic algorithms, machine learning, symbolic learning, supervised learning

1. Introduction

With the growing amount of available information, limited knowledge-extraction capabilities become a major bottleneck in understanding the world. Recognizing this, there has been an explosion of machine learning attempts to solve this problem. A big part of that research is devoted toward restricted attribute-based spaces. This can be attributed to the existence of many practical problems without a sufficiently understood body of knowledge, but with widely available data in the form of feature descriptions. Traditionally, all approaches have been classified as non-symbolic and symbolic, depending on the output language. Non-symbolic inductive learning systems, often called subsymbolic, usually do not acquire any explicit knowledge but rather gather other information necessary for the descriptive process. They include statistical models, where the only representation is by means of all stored examples or some statistics on them, and the connectionist models, where the knowledge is distributed among network connections and an activation method. Symbolic systems, on the other hand, produce explicit knowledge in a high-level descriptive language. However, an equally important distinction can be based on the level of inference. All non-symbolic approaches process low-level entities (usually numerical parameters). On the other hand, the level of inference of symbolic approaches widely varies, with those operating at higher level showing an advantage: mechanisms based on symbol manipulations, in addition to being closely related to their task objectives, allow for the use of the same input and output language. This, in turn, creates the possibility of employing some more sophisticated learning paradigms, such as incremental and closed-loop learning, and defining the processing on the same level, which is one of the most important features of our approach described in this article.

With the development of the computer, there has been an increasing interest in simulating nature as a means for problem solving. One of the best-known frameworks was developed by Holland (1975) and is known as genetic algorithms (GAs). This approach models the natural processes of inheritance of coded knowledge and survival by fitness or degree of adaptation to the environment. The two most important characteristics of GAs are robustness and the domain independence of their search mechanism. Robustness, an ultimate goal of any system, is a natural by-product of the search strategy, which performs simultaneous exploration and exploitation. This strategy makes the mechanism quite independent of the characteristics that normally cause difficulty for most other approaches, such as non-smoothness or discontinuity of the evaluation function. Domain independence, on the other hand, is obtained by designing the search operators in the space of the lowest-level representation. However, such an approach has both advantages and disadvantages. On the positive side is the fact that a new application requires only a coding of the problem to this artificial space. On the negative side lies the fact that the quality of such a coding is crucial to the genetic algorithm's performance. Moreover, operating in this space means using problem-blind operators that often overlook some important information that could be utilized to guide the search.

Nevertheless, genetic algorithms have been quite successfully applied to a number of problems. The most outstanding results come from the field of parameter optimization (DeJong, 1988), where the coding is rather straightforward. Other successful applications include optimization problems like wire routing and scheduling, game-playing, cognitive modeling, transportation problems, the traveling salesman problem, and control problems (e.g., Davis, 1991; DeJong, 1985; Goldberg, 1989; Grefenstette, 1987). However, applications to machine learning, although partially successful (Koza, 1989; Rendell, 1985; Schaffer, 1985; Sedbrook, Wright, & Wright, 1991), generally are still too weak for more complex domains and face many related issues (DeJong, 1988, 1990). The genetic algorithm approach to supervised learning in an attribute-based space is normally referred to as symbolic. However, the processing has traditionally been done in symbols of the artificial, not the problem, language. This has begun to change (for example, Grefenstette, 1991; Koza, 1989), following recent changes in the representation utilized (Spears & DeJong, 1990; Goldberg, 1985; Grefenstette, 1991; Smith, 1980).

Following this trend, we propose a different approach: rather than trying to extend the set of operators, we start with a set of inference rules specific to the task, and incorporate them into the genetic algorithm framework. To achieve this, we propose to use a rule-based representation as the natural choice for a symbolic system operating in this space. Having that, we can directly use the inference rules defined in such a framework, namely, those of the inductive learning methodology (Michalski, 1983). By doing so, we utilize the task-specific problem-solving methodology and abstract the genetic algorithm's inference to the problem-specific symbol level. This can be viewed as a knowledge-intensive approach, one using a vast amount of task-specific information, which replaces the blind search of traditional domain-independent operators by a heuristic search. Implementing all the extra knowledge in the operators leaves the remainder of the genetic algorithm intact and allows for a clear separation of specific knowledge from general mechanisms. Because of the richness of such new operators and their problem-specific behavior, the new algorithm does not enjoy the same theoretical foundations as the traditional GAs do. Nevertheless, we try to justify it intuitively, and the results of our initial experiments indicate its applicability.

Compared to the classical symbolic systems ID and AQ, this new approach, which combines the inductive inference rules with the quality of the GA search, leads to a potentially very robust design that does not assume any prior relationships among different attributes (as, e.g., ID does). The robustness is a result of the existence of the platform for both cooperation (by information exchange) and competition (by selection) among many different simultaneous solutions. This, in turn, can be seen as an extension of the AQ's ideas of processing competing directions (AQ does it by simultaneously retaining a number of partial covers). Here, we provide the cooperation, and we use more powerful heuristics: the inference rules and their adjusting applicabilities.

Successful results may find manifold applications. First of all, the system would provide a quality alternative to existing machine learning techniques. Moreover, applications of all competition and cooperation, generalization and specialization, and more powerful problem-specific heuristics should increase the system's potentials, especially in complex problems. Second, the system would provide a valuable tool to study inductive learning methodology and its inference methods, since the system's principles of operations are based on cooperation and competition among simultaneous applications of different inference rules of the methodology. Third, the approach explored here, which abstracts the genetic algorithm to provide a setting for an ease of accommodation of deep problem understanding to build a hybrid system, can easily be applied to other domains and tasks. Finally, these ideas provide a framework for modular separation of knowledge and performance components in such hybrid genetic algorithms. Such a separation proved to be very successful and desired in other AI architectures.

In this article we attempt to justify our approach, present the system's design, and conclude with some initial experimental results. More experimentation, aimed both at more comprehensive comparison of this system to others and at measuring its exact characteristics, is left to be completed in the near future and will be presented separately. This article is organized as follows. In section 2 we start with a brief description of the genetic algorithm framework used, and follow in section 3 with a specific problem attacked along with a discussion of related issues that reflect on the system's design. We follow with a brief overview of the previous approaches in section 4. Then, we describe our new ideas and their application to the design (section 5) in the simpler case of single concepts. We address some important implementation issues in section 6. In section 7 we attempt both to illustrate the system's behavior by tracing its applications and to present results of some initial experiments. Finally, in section 8, we draw some conclusions and outline work to be done in the future.

2. Genetic algorithms

Genetic algorithms are adaptive methods of searching a solution space by applying operators modeled after the natural genetic inheritance and simulating the Darwinian struggle for survival. They belong to the class of probabilistic algorithms, yet are distinguished by their different search method and relative insensitivity to local traps. In general, a GA performs a multi-directional search, and it encourages information formation and exchange among such directions. It does so by maintaining a population of proposed solutions (chromosomes) for a given problem. Each solution is represented in a fixed alphabet (often binary) with an established meaning. The population undergoes a simulated evolution: relatively "good" solutions produce offspring, which subsequently replace the "worse" ones. The estimate of the quality of a solution is based on an evaluation function, which plays the role of an environment. The existence of such a population provides for the superiority of genetic algorithms over pure hill-climbing methods, for at any time the GA provides for both exploitation of the most promising solutions and exploration of the search space. The simulation cycle is performed in three basic steps. During the selection step a new population is formed from stochastically best samples (with replacement). Then, some of the members of the newly selected populations recombine. Finally, all new individuals are reevaluated.

The mating process (recombination) is based on the application of two operators: mutation and crossover. Mutation introduces random varability into the population, and crossover exchanges random pieces of two chromosomes in the hope of propagating partial solutions. Because both of these operators are often defined on syntactic pieces of the underlying representation (when each chromosome is viewed as a sequence of the symbols of the low-level alphabet), the search has domain-independent properties. However, the applicability of a GA to a particular problem depends on the representation emphasizing meaningful semantic pieces of information (called building blocks) to be used by crossover operators. Consequently, this applicability may be reduced by operating on the low-level syntactic structures.

Specifying a genetic algorithm for a particular problem involves describing a number of components. Among them, the most important are a genetic representation for potential solutions to the problem, which also defines the search space of the algorithm; a method of generating the initial population of potential solutions; an evaluation function that plays the role of the environment, rating solutions in terms of their "fitness" or "adaptation" to this environment; genetic operators that alter the composition of chromosomes during recombination; and values for various parameters that the GA uses.

The theoretical foundations of genetic algorithms rely on the notion of a schema (e.g., Holland, 1975)—a similarity template allowing an exploration of similarities among chromosomes. Using schemata, a growth equation may be derived that indicates the following hypotheses (e.g., Goldberg, 1989): a genetic algorithm seeks near optimal performance through juxtaposition of special kinds of schemata—the building blocks. Although some effort has been made to prove this hypotheses, for most nontrivial applications we rely on empirical results. Nevertheless, this hypothesis suggests that the coding problem for a genetic algorithm is critical for its performance, and that such a coding should emphasize meaningful building blocks. This, in turn, suggests the following intuitive approach to problem solving by genetic algorithms: the problem representation in a GA should be such that conceptually related alleles are close together in the resulting genotype, where the closeness is defined relative to the crossover operators.

3. Inductive learning from examples

Concept learning is a fundamental cognitive process that involves learning descriptions of some categories of objects. When the objects are described by features (attribute-value pairs) based on a number of multi-valued attributes, the learning is said to be in an attribute-based space. A priori knowledge consists of a set of events that are examples of the space. When each such event is preclassified as belonging to a category, the learning is said to be supervised. The task is to generalize the a priori knowledge in order to produce descriptions of the concepts. When a rule-based framework is used to express the descriptions, the acquired knowledge is often called decision rules. Such rules can subsequently be used both to infer properties of the corresponding classes and to classify other, previously unclassified, events from the space.

The idea of a concept itself may be defined in many different ways, depending on the assumed concept representation and on methods of instance classification. In addition to psychological evidence, this choice is often dictated by the underlying knowledge acquisition method, or rather its output language. For example, assuming a sufficient set of attributes for the events to be consistent and a crisp view, a decision tree can naturally produce complete and consistent partition of the search space. This is so because the decision tree mechanism starts with the whole space and recursively cuts it into disjoint subspaces. On the other hand, it is more difficult for a set of rules to cover the search space in the same manner. Therefore, an extra mechanism is needed to account for possible cases of no-match and multiple-match when recognizing new events. Such problems can be avoided while learning single concepts if the system learns only the concept description and assumes that the subspace not covered by this description represents the complement of the concept. This simplified case is used in this initial design.

An important issue is that of defining both the input and the output language. The input language serves as an interface between the environment (the teacher) and the system. Therefore, it should combine requirements of both of these entities. Moreover, it should minimize inconsistencies among data. The output language serves as an interface between the system and the application environment. Therefore, it should combine the requirements of the learning system with those of the environment. For example, for a pure classification application, there is no need to express the acquired knowledge on a comprehensible level. The output interface is only to provide classifications of some new events. On the other hand, a learning agent used as a part of an intelligent system must be able to communicate its knowledge to other parts of the whole system. If such a system contains elements operating in a high-level language (as an expert system, human expert, etc.), our learning agent should be able to express its knowledge at the same level. Another reason, pointed out by Michalski (1986), relates to the increasing dependence on any automatically generated knowledge:

> An important implication . . . is that any new knowledge generated by machines should be subjected to close *human scrutiny* before it is used. This suggests an important goal for research in machine learning: if people have to understand and validate machine-generated knowledge, then machine learning systems should be equipped with adequate *explanation capabilities*. Furthermore, knowledge created by machines should be

expressed in forms closely corresponding to human descriptions and mental models of this knowledge; that is, such knowledge should satisfy what the author calls the *comprehensibility principle*.

One widely used language, which is closely associated with rules, is VL_1 (Michalski, Mozetic, Hong, & Lavrac, 1986). Variables (attributes) are the basic units having multi-valued domains. According to the relationship among different domain values, such domains may be of different types: nominal, for example $\{Yes, No\}$ in boolean attributes; linear with linearly ordered values; or structured with partially ordered values.

Relations "$=, \neq, <, \leq, >, \geq$" associate variables with their values by means of selectors having the form [*variable relation value*], with the natural semantics. For example, $[Age > Young]$ is interepreted as the set of people of *Middle* or *Old* age, assuming that the three values $\langle Young, Middle, Old \rangle$ are in the domain of the linear attribute *Age*. The *value* in a selector is a single domain value. However, for the "=" relation, it may be a disjunction of such values (the so-called internal disjunction). The internal disjunction, along with the natural semantics, makes the equality relation alone sufficient to represent any formula of the language. This important property is used in the design of our operators. Conjunctions of selectors form complexes.

The significance of the language relies on the natural correspondence to a rule-based paradigm. Selectors can be used to express conditions on single attributes. Complexes can be used to express rules of the form *complex ::> decision*. Because of that, the semantics of the language's constructs is easily understood. For example, the following set of rules describes people with heart problems as those who are older and have high blood pressure, or those with high cholesterol levels:

$$[BloodPressure = High][Age \neq Young] ::> HeartRiskGroup$$
$$[CholesterolLevel = High] ::> HeartRiskGroup$$

However, when only single concepts are being learned, the *decision* is redundant and can be omitted. Again, this property is used in our design.

There are two different approaches to learning from examples. The first assumes the existence of working memory able to remember all previously seen events for a future reference. This approach is normally referred to as full memory learning, and the incremental processing is associated with both previously generated knowledge and previously seen examples, in addition to the newly presented events (one should also mention here the case of so-called batch-incremental systems, which process the incrementally available events in relation to only the previously seen events, disregarding the previously generated knowledge). The feasibility of such full-memory systems is restricted by the data set size. However, the other advantages of these systems, especially their relative conceptual simplicity, cause many learning systems to follow this direction—as does our system. In addition, two other important factors favor this approach in the domain of attribute-based spaces: the available data sets for many interesting concepts are appropriately small, and such systems can normally accommodate the large data sets by means of some preprocessing mechanisms. The other approach assumes that the only available memory is for the generated body of knowledge.

Incremental learning capabilities are important characteristics of a learning system. This is so mainly for two reasons: human learning shows incremental characteristics, and a natural setting for a learning system often displays dynamic properties—new evidence becomes available from time to time, as might be the case in a medical database environment where some of the diagnoses may be confirmed after a while, or when new patients are diagnosed. Conceptually, the incremental character is obtained by abilities to generalize and specialize existing knowledge, as necessary, upon new experience. Since our design is based on the inductive learning methodology, which is incremental in nature, the system should possess such capabilities.

Michalski (1983) provides a detailed description of various inductive operators that constitute the process of inductive inference. In the restricted language VL_1 (for induction in an attribute-based space), the most important are as follows: *condition dropping*—that is, dropping a selector from the concept description; *adding alternative rule* and *dropping a rule*—adding/removing one rule from the description; *extending a reference*—extending an internal disjunction; *closing an interval*—for *linear* domains filling up missing values between two present values in a selector; *climbing generalization*—for structured domains climbing the generalization tree; *turning a conjunction to a disjunction*; *inductive resolution*—analogous to the resolution principle. These operators are either generalizing or specializing existing knowledge. There is no provision for generating the initial set of rules. In section 5 we define our versions of these operators to be used in our genetic learning system. We also discuss the choice of the initial knowledge.

4. Previous approaches

Over the past few decades there have been many different approaches to the problem of supervised learning in attribute-based spaces. Some of these approaches came from the AI community, and others from fields such as statistics. One of the most recent ideas has been to use genetic algorithms, to which we pay special attention since our new approach tries to extend such ideas by those of the inductive methodology.

4.1. Traditional approaches

As mentioned earlier, non-symbolic systems rely mostly on quantitative information processing. Therefore, they are further away from mainstream AI devoted to symbol processing. On the other hand, the symbolic systems apply the qualitative approach to learning: the output is a high-level description, and the processing itself is often done at the symbol level.

Statistical approaches account for the vast majority of non-symbolic, or numerical, approaches. They usually operate in batch mode on the data set in order to obtain some statistical measures, which are later used as probabilistic approximations of appearances of different features. To achieve a suitable classification, the correlation of this information to a new example is accumulated using some inference methods. Among such methods, the Bayesian probabilistic model is the best known. The disadvantage of these approaches is that they rely on low-level processing for high-level learning. Furthermore, such treatment makes it hard to process any available problem-specific knowledge. In addition, the measures used treat all features independently, and high processing complexity does not allow exploration of inter-feature dependencies, even though they could be incorporated (Rendell, Cho, & Seshu, 1989).

Another numerical approach comes from the neural network community. A neural network is a cognitive model of the brain and is composed of two kinds of elements: processing elements (nodes of the network) and connections. Viewed as a memory, such a network has its knowledge distributed among the connections—called weights. These weights determine the propagation of excitatory and inhibitory signals that, in turn, determine the excitation of certain nodes. Such a memory model is capable of learning. The backward propagation of a failure is the best-known method of setting the weights. A neural network method has been applied to simple cases of concept learning with some successes. However, these applications are usually quantitative as well, which makes it difficult to establish a platform for any higher-level knowledge utilization or understanding.

As described, the non-symbolic systems do not follow the methods of the inductive learning methodology, but rather perform numerical computations. This, however, leads to an apparent advantage of better applicability to processing noisy information (Rendel et al., 1989) and more gradual performance degradation.

The two prominent symbolic approaches to supervised feature-based learning, recognized as benchmarks, are Michalski's AQ (Michalski, 1983) and Quinlan's ID (Quinlan, 1986). They are both considerd symbolic systems, even though they have some numerical elements: ID uses an information measure function, while the two-tiered representation of AQ performs a partially probabilistic inference.

The AQ approach is based on inductive generalization and specialization of the VL_1 formulas using the idea of a cover of the positive against the negative events. The cover is constructed in an iterative manner, starting with only one positive and one negative event and continuing until the generated cover is complete and consistent. To prevent an apparent exponential growth in the number of generated descriptions, special heuristics, which accommodate user-defined learning criteria, are employed to reduce the size of partial covers. Retaining a number of such current covers provides for a competition among different solutions. This approach conceptually follows the ideas of inductive methodology, since the generated knowledge is either generalized or specialized, as appropriate. However, the algorithm itself uses only the logic-based operators of negation, union, and intersection to process the current descriptions.

In the ID approach, the training examples are represented by feature vectors similar to events in VL_1. The algorithm constructs a decision tree, where each leaf is associated with a single decision class and each internal node corresponds to an attribute, while each node's branches correspond to a value of that attribute. One of the features of such a tree is that no path from the root to a leaf has two nodes corresponding to the same attribute. The algorithm itself is an iterative application of the information content formula: $I = p * log(p)$, where p is a probability of given information (Quinlan, 1986). At each node of the tree the algorithm only treats events satisfied by the path to this node: the information content is calculated for all such remaining attributes and events, and the attribute giving the maximal information gain is selected as the label for this node. This approach, despite its apparent assumption of attribute independence, proved to be successful in terms of recognition quality. However, the numerical formula used causes serious problems when one tries to incorporate some task-specific knowledge. Moreover, the algorithm is conceptually very distant from the inductive learning methodology. It iteratively applies specialization, starting with the whole event space. Only then generalization can be applied, by means of tree pruning or rule construction techniques.

Both of the above are full-memory systems, meaning that they assume the availability of all previously seen events at any time during incremental learning. However, they both allow for processing large quantities of data: the AQ uses a preprocessing mechanism selecting only the most representative events, and the ID uses the idea of data windowing.

4.2. Genetic algorithm approaches

Since the early 1980s, there has been an increasing interest in applying GA methods to machine learning—in particular to learning production rules, whose special case is the problem of supervised learning from examples of an attribute-based space. The main problem in such applications is to find a suitable representation, able both to capture the desired problem characteristics and to represent a potential solution. Using a rule-based concept representation brings a different kind of problem: the number of such rules (disjuncts) is not known a priori. Therefore, the traditional fixed-length representation is unsuitable. Two different approaches have been proposed:

- **Michigan** approach, where the population still consists of fixed-length elements, but the solution is represented by a set of chromosomes from the population. This methodology, known as CS for classifier systems, along with a special "bucket brigade" mechanism for credit assignment, was originally developed by Holland and colleagues (Holland, 1986). Here, each chromosome, called a classifier, represents a structure composed of conditions and messages lists. The environment, together with the activated rules, provides a set of active messages. These, in turn, activate other classifiers by satisfying their conditions. The chained actions of message–condition pairs cluster the rules together. Because of this chaining mechanism, this approach seems more suitable for planning than concept learning.
- **Pittsburgh** approach, which represents an extension of the traditional fixed-length chromosome approaches. Here, variable-length chromosomes are used to represent proposed solutions individually. This approach (LS for Learning System) was originally proposed by Smith (1980). This representation seems more naturally suited for the supervised learning, since each chromosome represents an independent solution.

Both of these approaches suffer from some drawbacks. A chromosome of a classifier system does not individually represent a solution. The variable-length approach constitutes a wide divergence from the traditional GA, and, therefore, requires special treatment. Nevertheless, some applications prove to be successful, although in quite limited applications. The two most noticeable genetic algorithm approaches to supervised concept learning in attribute-based spaces, in the LS framework, come from Koza and Spears with DeJong. Koza uses Lisp programs as a means of representing potential solutions, with some tree-based operators that are closed in the space of such representation. Spears and DeJong use a binary representation for multi-valued domains, and implement only the traditional operators of mutation and crossover in their GABIL system (Spears & DeJong, 1990). Another important LS system is SAMUEL (Grefenstette, 1991), but it was designed and applied mostly to sequential decision problems. Classifier systems have also been used for concept learning (e.g., Wilson, 1987). However there are some additional problems associated with the indirect solution (DeJong, 1990). Moreover, as pointed out in Liepins and Wang (1991), the traditional CS architecture is not suitable for stationary concept learning. Nevertheless, some recent modifications relax these limitations (Booker, 1989).

With very few exceptions (e.g., Spears & DeJong, 1990; Grefenstette, 1991; Koza, 1989), all systems for rule-based learning use a three-symbol alphabet {0, 1, #}, where # stands for a wild-card character (e.g., Goldberg, 1985; Greene & Smith, 1987; Schaffer, 1985), to code each individual feature. Such an alphabet is the choice of many CS- and LS-based systems. However, it is not so well suited for non-binary domains, since it often increases the representation length and extends the search space to infeasible regions. Then an additional mechanism must be implemented to deal with these problems (Green & Smith, 1987). A more systematic approach would be to model closely the non-redundant and feasible-only representation of VL_1, as used in this work and independently in Spears and DeJong (1990). The approach of Grefenstette (1991) also follows a similar direction.

5. The modified genetic algorithm

In this section we first describe the major ideas used in the system's design, and then present the system itself by defining the necessary GA components. Some implementation issues, strongly associated with any such approach, are addressed in the next section.

5.1. Ideas used

One of the most praised characteristics of a genetic algorithm is its domain-independent search (DeJong, 1988). This is the source both of the many successes of GAs, especially in parameter optimization, and, at the same time, of many limitations in other applications. In general, such arguments here are similar to those calling for more task-specific AI methods two decades ago. Recall that, at first, general problem solvers were devised, which were to play the role of general tools for problem solving. It soon turned out that, due mostly to the unmanageable complexity of such methods, it was necessary for the designers of intelligent systems to incorporate domain- and problem-specific knowledge, either by making it explicit or by hiding it in the implementation.

The need for problem-specific knowledge incorporation into GAs was recognized as a method for improvement in many different domains (e.g., Grefenstette, 1987). Davis (1991) calls for such approaches where he calls them "hybrid" genetic algorithms and argues for the exploration of combinations of GAs with existing methodologies in any possible domain. Similar ideas were called for in applications to machine learning. For example, Forrest (1985) proposed using high-level operators such as "concept specialization" and

Table 1. The spectrum of knowledge incorporation in a GA.

Level of task-knowledge incorporated	
None	Full
Only classical mutation and crossover operators	No domain-independent operators, only fully implemented problem-solving methodology

"value restriction"; Antonisse and Keller (1987) also called for similar incorporations. The above were restricted to classifier system architectures. In the LS approach, Grefenstette's SAMUEL system (Grefenstette, 1991) used similar problem-specific operators, including specialization, generalization, rule merge, and delete. However, that approach was aimed at more general sequential decision problems and non-full-memory learning.

Following the Pittsburgh approach, which allows direct representations of VL_1 solutions, we are faced with chromosomes of varying number of structures. Then there is a whole spectrum of possible GA designs along the dimension of task-specific knowledge utilization (see table 1). On one side of the spectrum lies a method that only uses the classicial operators of mutation and crossover. This approach is conceptually very easy. Moreover, it enjoys the same theoretical foundations as the fixed-length GAs: Smith (1980) showed it was true provided that such structures are positionally independent. The above determines the existing popularity of such approaches in any domain. The same is true in the particular case of supervised inductive learning. For example, the previously mentioned GABIL system follows this path.

On the other side of the spectrum lies a knowledge-intensive method that completely abandons the traditional domain-independent operators, and, instead, fully implements the specific problem-solving methodology. This approach is conceptually much more challenging, since it requires, in addition to the GA implementation, a clear understanding of the problem being solved, along with a well-described, complete solving methodology defined at the problem level. This fact, in addition to the lack of well-established theoretical foundations and the resulting diversion from models of nature, determines the low popularity of such approaches. Nevertheless, some machine learning approaches drift in this direction. For example, the new Lamarckian learning operators used in SAMUEL (Grefenstette, 1991) implement learning-specific mutation and crossover.

Even though this latter approach does not have the same theoretical support, it is backed by the task-specific knowledge used to guide and conduct the search. This property should provide for a faster convergence to a desired solution. Moreover, it may be easily shown that all the operators we subsequently define and use (section 5.5) are actually special cases of the traditional mutation and crossover. This provides an intuitive support for the same theoretical foundations. Also, because the operators are defined on the semantic pieces of the problem, one may easily argue that this design naturally satisfies the building-block hypothesis as well.

These are some of the reasons for our decision to pursue this path. But an even more appealing and important justification arises in the context of the learning process understanding and validation. Applications of the task-specific knowledge as the means for inference

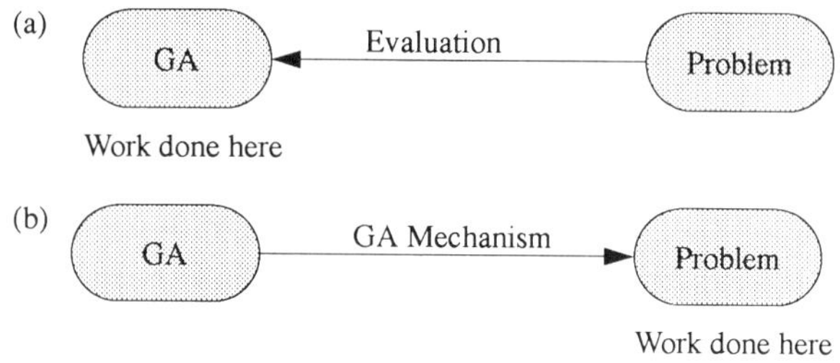

Figure 1. A GA with (a) traditional and (b) task-specific operators.

mechanisms provide for a better understanding of the underlying principles of the learning system. This becomes increasingly important while designing systems that are not only able to generate knowledge, but also able to explain and justify their behavior. For example, Michalski wrote (Michalski et al., 1986):

> . . . one should strive to facilitate human understanding not only of the surface results but also of the underlying principles, assumptions, and theories that lead to these results.

This approach is also justified as an abstraction of the genetic algorithm approach. Following the previous discussion on GAs, and those intuitive results stating that the best representation should provide the chromosome structure reflecting syntactic and conceptual knowledge of the problem, we actually go to the extreme of using the problem space as the working search space. In other words, while applications of the traditional domain-independent operators provide for a domain-independent search conducted in the artificial representation space (figure 1a), we set the genetic algorithm to operate directly in the problem space by organizing the work there (figure 1b).

5.2. Representation and search space

We adopt the multiple-valued logic language VL_1 as the choice for chromosome's representation. Then the search space is the space of sets of rules, spanned by given features. This is the space of VL_1 concept descriptions. Because we do not employ any extra axioms, it is quite feasible and possible to have redundant descriptions, such as

$$[Age > Young] ::> HeartRiskGroup,\ [Age = Old] ::> HeartRiskGroup$$

For simplicity of presentation (but not lack of generality), from now on we only consider VL_1 formulas built using the sufficient "=" relation with internal disjunctions. Moreover, for the same reasons, we assume that we are dealing with single concepts, and that we are learning only a single description (as, e.g., in Spears & DeJong, 1990). This simplification allows us to assume a crisp rule-based conceptualization. We address possible generalizations of this approach to multiple concepts and non-crisp views in the part on future research directions (section 8).

Because of the assumption of learning only a single concept description, all rules are associated with the same single decision, which subsequently does not have to be stated explicitly. Accordingly, when no confusion can arise, we may refer to the same set of rules as just a logical disjunction of VL_1 complexes.

5.3. Initial population

The population contains individuals, each of which is a potentially feasible solution (a set of rules of the VL_1 language). Its size remains fixed (as a parameter of the system). Initially the population must be filled with potential solutions. This process might be totally random (as is normally the case in genetic algorithms), or it might incorporate some task-specific knowledge. There is an obvious trade-off between the level of knowledge used in such an intelligent initialization. On one side of the spectrum is the random choice, very cheap and simple. On the other side, we have an initialization that produces actual solutions to the problem, differing possibly by some applied criteria. This later initialization is actually as hard as the problem we wish to solve. Therefore, it is inapplicable.

We follow the idea of an initialization that is as simple as possible, yet intelligent. Accordingly, we allow for three different types of chromosomes to fill the population initially:

1. The first type is a random initialization. Each individual is a set of a random number of complexes, randomly generated on the search space.
2. The second type is initialization with data. Each individual is a random positive training event.
3. The third type is initialization with prior hypotheses, provided such are available. Each individual is just a single hypothesis given a priori. Having such capabilities, the system can be used as a knowledge refinement tool—possibly cooperating with an expert system of an intelligent hybrid framework.

Actual experiments show that the best average behavior is obtained while using a combination of these three (or the first two if initial hypotheses are not specified), even though the importance of the initialization with positive events seems to diminish with the use of an operator that adds such positive events to current descriptions.

5.4. Evaluation mechanism

The evaluation function must reflect the learning criteria. In supervised learning from examples, the criteria normally include completeness, consistency, and possibly complexity. In general, one may wish to accommodate some additional criteria, such as cost of attributes, length of descriptions, their generality, etc., but we did not consider them in the current implementation.

The completeness and consistency of a rule, or a rule set, measures its quality with respect to the set of training events. We use the formulas presented in table 2, where e^+/e^- is the number of positive/negative training events currently covered by a rule, ϵ^+/ϵ^- is the

Table 2. Completeness and consistency measures.

Structure Type	Completeness	Consistency
A rule set	ϵ^+/E^+	$1 - \epsilon^-/E^-$
A rule	e^+/ϵ^+	$1 - e^-/\epsilon^-$

number of such events covered by a rule set, and E^+/E^- is the total number of such events. These two measures are meaningful only to rule sets and individual rules. For conditions, the measures of the parent rule are used. These definitions assume the full-memory model.

Combining multiple criteria in a single evaluation measure is very dificult and critical for the convergence problem (Goldberg, 1989). In our case we need to combine three such values. We can ease this task by replacing the completeness and consistency measures with a single measure of correctness:

$$correctness = \frac{w_1 \cdot completeness + w_2 \cdot consistency}{w_1 + w_2}$$

This is only one of possible combinations, and in the future we plan to explore the choice's dependence on some problem-specific meta-level knowledge. Finally, we combine correctness and cost by

$$evaluation = correctness \cdot (1 + w_3 \cdot (1 - cost))^f$$

where w_3 determines the influence of *cost* (which itself is normalized on [0, 1]), and f grows very slowly on [0, 1] as the population ages (a dynamic approach). The description's cost is measured by its complexity: *complexity* = 2 · *#rules* + *#conditions*, as in Wnek, Sarma, Wahab, and Michalski (1990). The above evaluation function is an experimental rather than a theoretical choice, and provides for a controlled bias with respect to complexity of descriptions. Moreover, the w_1 and w_2 coefficients bias the search to more complete or consistent descriptions.

For most practical tests we used a very low w_3 weight ($\sim$ 0.01). Too high a value may cause weak rules to be dropped from the descriptions; too low a value reduces the probability of simplifying the generated descriptions (e.g., dropping redundant rules). The primary reason for the cost accommodation is to force differentiation between the same or similarly covering rule sets that have different complexity. We use a dynamic approach to the use of cost (an approach that adjusts its effects as the population ages). We successfully applied similar dynamic ideas in other domains (e.g., Michalewicz & Janikow, 1991). The effect of the very slowly raising f is that initially the cost influence is very light in order to promote wider space exploration, and it only increases at later stages in order to minimize complexity. Moreover, initial experiments suggest that the system performs better when the f exponent somehow fluctuates, and that the final increase should start based upon an anticipated exhaustion of resources or when the currently learned description is already complete and consistent.

5.5. Operators

The operators transform chromosomes to new (possibly better) states in the search space. Since the system operates in the problem space, the operators directly follow the inductive learning methodology. Accordingly to the three syntactic levels of the rule-based framework (conditions, rules, rule sets), we divide the operators into three corresponding groups. In addition, each operator is classified as having either generalizing, specializing, or unspecified—or independent—behaviors. Note that the inductive methodology does not define independent operators. Their introduction is strongly associated with the use of a poulation.

For a better illustration, we exemplify some of the operators using the following search space:

Attribute	Values	Type
A	0, 1, 2, 3	Linear
B	0, 1	Nominal
C	0, 1, 2	Nominal
D	0, 1	Nominal

and the idea of diagramatic visualization (Wnek et al., 1990), which is a multi-valued extension of the well known "Karnaugh map" or "Veitch diagram." Following the correspondence of VL_1 complexes and rules in the single-concept scenario, we represent a rule set as a disjunction of VL_1 complexes. Each complex is a left-hand side of a rule corresponding to the same decision. Also, we try to define the operators as simply as possible, in order to reduce the computational overhead. For example, we do not use the inductive resolution rule, which requires an extensive pattern matching in order to find two complexes having selectors that are negations of each other.

5.5.1. Rule set level

This is the level of sets of VL_1 complexes. Here, the operators act on whole rule sets (one or two at a time):

- Independent:

  Rules exchange. This operator requires two parent rule sets, and it exchanges random rules between these two. It requires two parameters: probability of application to a rule set, and probability of rules selection for the exchange.
- Generalization:

  Rules copy. This operator requires two parent rule sets, and it copies a random rule from each of the sets to the other. It differs from the "rules exchange" operator, since it does not remove information being propagated from the rule set. It requires one parameter: probability of application to a rule set.

  New event. This operator acts on a single rule set: if there is a positive event not covered yet by the current rule set, this event's description is added to the set as a new rule:

$$chrom = \bigcup_i (cpx_i ::> dec) \text{ and } \exists_{e^+} \forall_i (e^+ \not\Rightarrow cpx_i) \lhd chrom \cup (e^+ ::> dec)$$

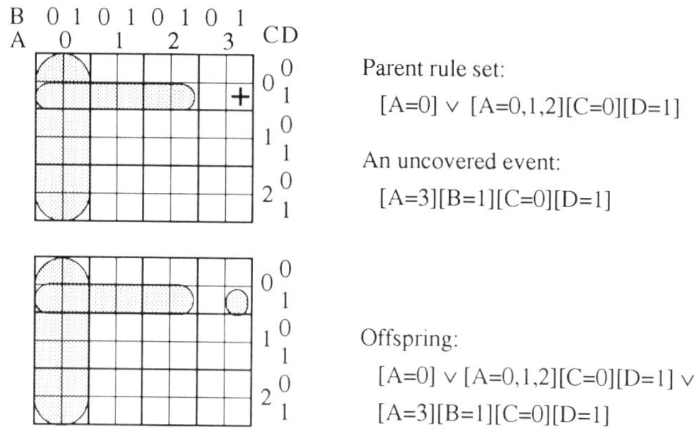

Figure 2. A visualization of the "new event" operator.

It requires only one parameter: probability of application to a rule set. Contrary to expectations, there is no pattern matching involved in this operator, and its actions have very little computational overhead due to data compilation—see section 6.3. For an illustration, see figure 2.

Rules generalization. This operator acts on a single rule set. It selects two random rules and replaces them by their most specific generalization:

$$cpx_1 ::> dec,\ cpx_2 ::> dec \lhd (cpx' ::> dec)$$

where *cpx'* is the most specific generalization (not necessarily consistent with respect to previously excluded negative events) of the two complexes. It requires one parameter: probability of application to a rule set. For an illustration, see figure 3.

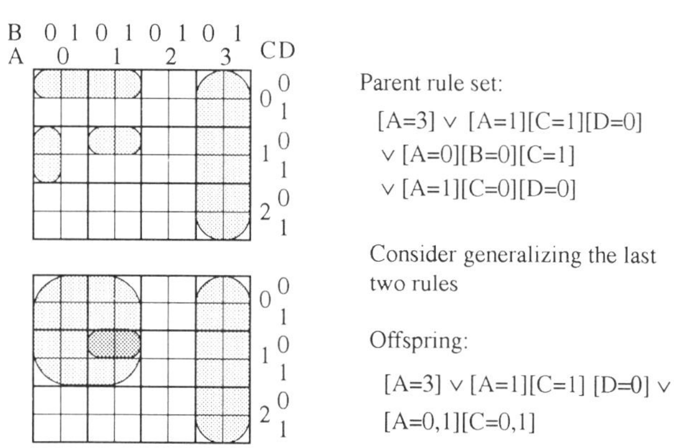

Figure 3. A visualization of the "rules generalization" operator.

- Specialization:
  Rules drop. This operator acts on a single rule set, and it drops a random rule from that set. It requires one parameter: probability of application to a rule.
  Rules specialization. This operator acts on a single rule set, and it replaces two random rules by their most general specialization:

$$cpx_1 ::> dec,\ cpx_2 ::> dec \ \triangleright\ (cpx' ::> dec)$$

  where cpx' is the most general specialization (not necessarily complete) of the two complexes. It requires one parameter: probability of application to a rule set.

5.5.2. Rule level

This is the level of VL_1 complexes. Here, the operators act on one rule at a time.

- Independent:
  Rule split. This operator acts on a single rule, and it splits it into a number of rules, according to a partition of the values of a condition (an absent condition can be selected as well, using all domain values). The set of domain values present in the selected condition can be split according to each value individually or according to two disjoint subsets of values. For the linear data types, the latter split is more desired. In this case, the present domain values are split by cutting the ordered set of values in a single random place. For the nominal data type, the former split is more desired. A structured type requires a slightly more sophisticated approach, similar to that of the linear type, but with differently defined values and orderings. This operator requires three parameters: probability of application to a rule, and probabilities of a subset vs. all values split, separately for the linear and nominal data types.
- Generalization:
  Condition drop. This operator acts on a single rule, and it removes a present condition from that rule:

$$((cpx = \wedge_i\ sel_i) ::> dec) \ \triangleleft\ (cpx' ::> dec)$$

  where cpx' has all but one selectors of cpx. In other words, one of the selectors of cpx is extended to cover the whole domain of the associated attribute. It requires a single parameter: probability of application to a rule.
  Turning conjunction into disjunction. This operator acts on a single rule, and it splits the complex into a disjunction:

$$((\wedge_i\ sel_i \wedge \wedge_j\ sel_j) ::> dec) \ \triangleleft\ (\wedge_i\ sel_i ::> dec) \vee (\wedge_j\ sel_j ::> dec)$$

  where the complex's separation into n and m selectors is random and position independent. It requires a single parameter: probability of application to a rule.

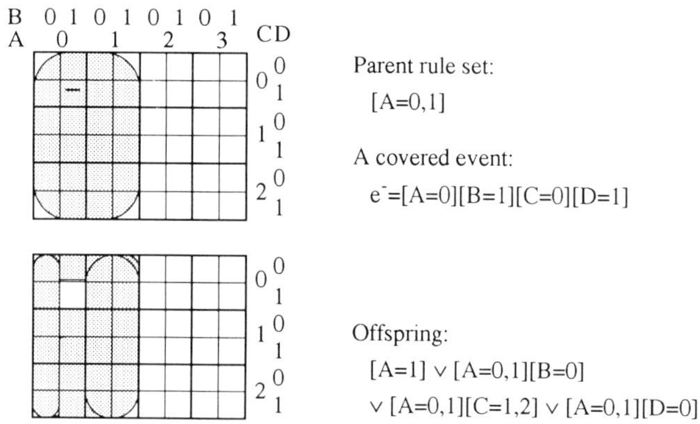

Figure 4. A visualization of the "rule directed split" operator.

- Specialization:

Condition introduce. This operator acts on a single rule, and it introduces a random condition associated with an unconditioned attribute:

$$((cpx = \wedge_i\ sel_i) ::> dec) \rhd (cpx' ::> dec)$$

where cpx' has, in addition to all the selectors of cpx, a new selector associated with an attribute not present in cpx. The new selector is a random choice from among all of its possible internal disjunctions. It requires a single parameter: probability of application to a rule.

Rule directed split. This operator acts on a single rule. If this rule covers a negative event, it is split into a set of maximally general rules that are yet consistent with that event, in the following way:

$$(cpx ::> dec) \text{ and } \exists_{e^-}(e^- \Rightarrow cpx) \rhd \cup_i(cpx_i ::> dec)$$

where the new set has cpx's such that $(\vee_i\ cpx_i) = (cpx \wedge \neg e^-)$. This operator resembles the action in the heart of the covered procedure of AQ15. It requires a single parameter: probability of application to a rule. Again, there is very little computational overhead involved due to data compilation. For an illustration, see figure 4.

5.5.3. Condition level

This is the level of VL_1 selectors. The operators act on one condition at a time.

- Independent

Reference change. This operator acts on a single condition, and it randomly removes or adds a single domain value to this condition. It requires a single parameter: probability of application to a condition.

- Generalization:
  Reference extension. This operator acts on a single condition, and it extends the domain by allowing a number of additional values. For the nominal type, some random values are selected for extension. For the linear type, a single value may be selected or, with a higher chance, a range may be closed (between two present values, using the domain ordering). Moreover, such shorter open ranges have a higher chance of being selected over longer ones. This operator requires quite a few parameters, including a probability of application to a condition, and a number of selection probabilities determining the choice of an action. For the structured type, we replace some of the present values by their parent in the generalization tree, giving preference to those offsprings that prevail in number (generalization climbing).
- Specialization:
  Reference restriction. This operator acts on a single condition, and it removes some domain values from this condition. Its actions and parameters are analogous to those of the "reference extension," but have opposite effects.

5.5.4. *Operator selection probability*

As seen in the definitions, each operator is given some initial probabilities from two separate groups: selection and application probabilities.

The selection probabilities serve as means of selecting one of a number of possible actions or substructures to participate in the operations—these probabilities are static. The application probabilities serve as a means of firing operators for structures of their type (based on the definition level). These probabilities have a dynamic character with respect to the current context, that is, to both the current coverage and the current, problem-dependent, size of the average chromosome. First, probabilities of generalizing operators are increased for applications to structures (rule sets, rules, conditions) that are incomplete, and are decreased for those that are inconsistent. On the other hand, the probabilities of specializing operators are increased for applications to structures that are inconsistent, and decreased for those that are incomplete. Moreover, the levels of probability increase/decrease are based on the levels of inconsistency/incompleteness. In other words, these two measures serve as additional heuristics (in addition to fitness), which guide the selection of appropriate operators. Second, all application probabilities are adjusted to achieve a constant chromosome update rate. For example, more complex problems, which cause the intermediate chrmosomes to be longer (both in terms of the number of complexes and their sizes), decrease all such probabilities by the same fraction.

While selecting the appropriate bias for completeness and consistency, we must be careful not to decrease the probabilities so far as to prevent certain operations from performing. Since we want the changes to be proportional to those measures, the following choices seem natural (but still experimental):

$$\text{Generalizing operators: } p' = p \cdot \left(\frac{3}{2} - \mathit{completeness}\right) \cdot \left(\frac{1}{2} + \mathit{consistency}\right)$$

$$\text{Specializing operators: } p' = p \cdot \left(\frac{1}{2} + \mathit{completeness}\right) \cdot \left(\frac{3}{2} - \mathit{consistency}\right)$$

The new value p' is the adjusted probability, and p is the actual probability. It is important to mention that since p' is computed differently for each rule and rule set, it does not replace the a priori p. The simplicity of this formulas guarantees a low computational overhead.

To accommodate the changes in problem-specific characteristics, namely, the average size of a complex and the average length of chromosomes in the current population, we use the following approximation. We observe the number of chromosomes undergoing recombination in a given population. If this number represents too large a portion of the population, we decrease all the a priori application probabilities by a fraction for the next reproductive iteration. If this number is too small, we do the opposite. Since the adjustments are computed for the whole population, they actually always replace the previous values. Experiments show that with an appropriately small such adjustment fraction, the changes converge and then the probabilities remain relatively steady. This method provides for a partial independence of such probabilities from some characteristics of the problems.

5.6. Algorithm

The algorithm uses the above components and the control of genetic algorithms. At each iteration, all rule sets of the population are evaluated and a new population is formed by drawing members from the original one in such a way that more fit individuals have a higher chance of being selected (sections 2 and 5.4). Following that, the operators are applied to population members in order to move these partial solutions, hopefully, closer to the desired state. Each structure stochastically invokes some operators defined for its level: the selection depends on the operators' initial strength, the consistency/completeness of the structure, and the average size of the current chromosomes. Then the cycle repeats until a desired description is found or computational resources are exhausted. For a better illustration, refer to the experimental tracing of section 7.2.

6. Some implementation issues

We implemented a simple **C** prototype of the proposed approach in order to be able to test our ideas. The choice of the language, aside from efficiency, was based on availability of bitwise logical operators used to speed up the evaluation mechanism (section 6.3). We call this implementation **GIL** (for **G**enetic-Based **I**nductive **L**earning). In this section we present the most important issues facing any such implementation, along with approaches used in GIL.

6.1. Sampling mechanism

The selection algorithm is to choose some chromosomes from the current population to form a new population, with possible omissions or repetitions. There are many standard ways of performing this step. Baker (1987) provides an excellent discussion. We use the stochastic universal sampling mechanism. This method builds a roulette wheel for the chromosomes, with each chromosome having allocated a portion of the wheel proportional to its evaluation fitness. A second wheel is constructed with equally spaced marks. The number of such marks is the same as the number of samples to be drawn (the size of the population). The wheels are placed on a single axis and the one with marks is randomly spun against the other. The positions of the marks, in relation to the space allocated for each chromosome on the other wheel, are then observed. A chromosome is selected once for each mark landing in its allocated space.

6.2. Internal representation

In section 5.2 we described the architecture of the chromosome, in terms of the language used. Now we discuss some important issues associated with the internal representation.

In section 4.2 we mentioned that the widely used three-symbol alphabet is not suitable for the multi-valued domains of feature-based spaces. A more appropriate solution was suggested originally by Greene and Smith (1987) and was recently used by Spears and DeJong in their GABIL system (Spears & DeJong, 1990). This approach uses binary digits to represent domain values. For example, assuming that an attribute has five domain values, the binary vector 11001 represents the condition saying that the attribute must have the first, second, or the last value (assuming some positional enumeration of values from the left, and a use of the internal disjunction). We use exactly these ideas to implement conditions (selectors of VL_1). A chromosome's length is actually unrestricted—a rule set may contain any number of rules, organized as linked lists. An important issue associated with the rule set is that of treating rules that are invalid, that is, conditions that are totally restricted (exclude all domain values). Spears and DeJong (1990) suggested keeping such rules as possible sources of valid conditions. We performed some experiments to study the trade-off between anticipated increases in the computational cost of retaining these rules vs. improvements in predictive accuracy of the system. Our conclusions are far from final. Nevertheless, they suggest there is a clear conflict between these two factors. A similar issue arises in the context of empty rules—those not covering any positive events. Such rules, again, may be removed or retained. Also, there seems to be a similar kind of complexity vs. quality trade-off. Currently, GIL removes both invalid and empty rules.

Each complex is a conjunction of a number of conditions. The number of such possible conditions, in a given complex, is bounded by the total number of attributes. We use that bound as a way of simplifying the internal representation: a complex is represented by a vector of conditions. Furthermore, for simplicity and efficiency, we associate a fixed positional correspondence between the attributes of the vector. This does not introduce any problems, since no operators acting at the condition level are positionally dependent.

Such an implementation introduces a new dilemma: how to treat unrestricted conditions, that is, those that include all domain values in the selector. It is a question of elegance, or possible efficiency, rather than power: an unrestricted selector can be dropped from its complex without any semantic change to the rule associated with this complex. We tried both approaches: one with all selectors present in each complex, and the other with

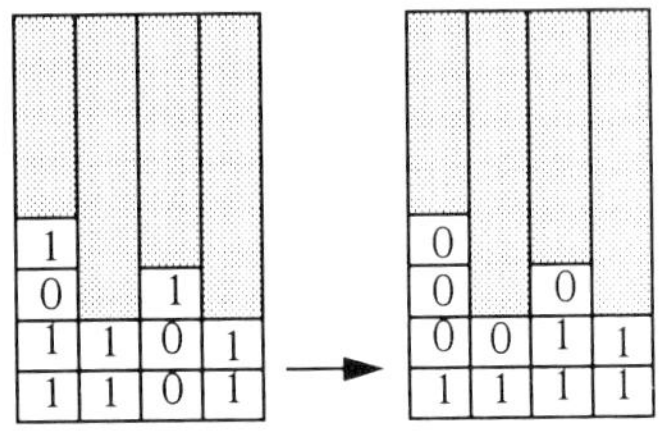

Figure 5. An internal representation of a sample chromosome.

unrestricted selectors invisible to all operators by means of a special flag. We found no significant difference in the system's performance under both conditions; we used the latter approach in GIL.

The above ideas are illustrated in figure 5, assuming the set of features used in section 5.5 while defining the operators on an eight-bit machine. This figure shows an internal representation of a chromosome, which can be viewed as a disjunction of complexes or a set of rules using an implicit decision. Moreover, depending on the treatment of unrestricted domains, the same chromosome can be described in different ways as well. The resulting four possible semantically equivalent views are as follows:

1. $[A = 0, 1, 3][B = 0, 1][C = 2][D = 0, 1]$ ::> *dec*, $[A = 0][B = 0][C = 0][D = 0, 1]$ ::> *dec*
2. $[A = 0, 1, 3][C = 2]$ ::> *dec*, $[A = 0][B = 0][C = 0, 1]$ ::> *dec*
3. $[A = 0, 1, 3][B = 0, 1][C = 2][D = 0, 1] \vee [A = 0][B = 0][C = 0, 1][D = 0, 1]$
4. $[A = 0, 1, 3][C = 2] \vee [A = 0][B = 0][C = 0, 1]$

6.3. Data compilation

A very commonly cited disadvantage of genetic approaches to problem solving is their time complexity (e.g., Quinlan, 1988). This problem becomes especially visible when the evaluation requires an extensive computation. This is the case in supervised learning when evaluating rule sets in a full-memory system, since this process involves extensive pattern matching. Concerned with such problems, we designed a special method of data compilation aimed at improving the time complexity of the system.

The idea is as follows: rather than storing data in terms of features, store features in terms of data coverage. In other words, for each possible feature, retain information about the events covered by this feature. This must be done separately for each concept. Moreover, it must be done even for concepts not being explicitly learned. For example, this means that GIL has to remember such coverage separately for both the concept and its negation. We achieve this by enumerating all learning events and constructing binary coverage vectors. This approach is feasible only for problems with small to medium size of data set. For those rare cases of larger sets (e.g., those larger than 10^4), special data-reduction mechanisms would have to be employed.

Table 3. Examples of binary coverage vectors of a feature.

Positive coverage vector:	1000000100000000101000000
Negative coverage vector:	000000000001010

The idea behind these vectors is analogous to that of representing conditions. A coverage vector is constructed for both E^+ and E^- separately. In this vector, a binary one at position n indicates that the structure that owns this coverage vector covers event $\#n$. For example, the vectors in table 3 indicate that the given feature covers positive events #1, 8, 17, 19 (out of 25), and negative events #12 and 14 (out of 15).

Prior to learning, as mentioned, all data are precompiled into such vectors, for all possible features. During the actual run of the system, similar vectors are constructed for all structures of the database: from the features upwards. For example, having the feature coverages, we can easily construct both positive and negative coverage of the condition $[A = 0, 2]$, by means of a simple bitwise OR on appropriate coverage vectors of features $(A = 0)$ and $(A = 2)$. Subsequently, condition coverages are propagated to rules by means of a simple bitwise AND. Finally, rule coverages are propagated to rule sets again by means of the bitwise OR.

Perhaps the most important effect of such an approach is that we can incrementally upgrade such coverages using a minimal amount of work after the initial database is fully covered. For example, consider a copy of the "rules copy" operator applied to the following two rule sets:

$$R_1 = r_1^1, r_1^2$$

$$R_2 = r_2^1, r_2^2, r_2^3$$

and suppose the operator copies r_2^2 to R_1. The coverage of the second rule set does not change. To compute the coverage of the first rule set, it is sufficient to perform bitwise OR between the coverage of the rule r_2^2 (which did not change during this operation) with the coverage of the original R_1. In other words, we compute this coverage using two bitwise OR operations (one for the positive and one for the negative coverage). In general, the number of such required operations increases (very slowly linearly) with the number of training events.

As another example, consider the case of the "reference change" operation, with a single change from 0 to 1, on position #, in a condition's binary vector. All that needs to be done to update the coverage of this condition is to perform bitwise OR on the coverages (positive and negative) of the corresponding feature number # associated with the given attribute with those of the original condition. Then this change must be propagated to the appropriate rule and rule set's coverages, using similar simple computations.

7. Experimental studies

In this section we attempt both to illustrate the system's behavior by tracing its applications and to experimentally compare its behavior to that of other learning methods. A more comprehensive testing to determine the system's characteristics is left for the future.

7.1. Experimental methodology

In the literature, learning systems are often evaluated and compared using a standard set of well-recognized artificial and real data. Examples widely used are random DNFs, multiplexers, soybean disease, breast cancer, etc. To evaluate GIL, we use some of these standard data sets. This also allows us to use published results of experiments with other systems, under the assumption of repeating exactly the same experimental sessions.

In inductive learning, the acquired knowledge should meet two criteria: high predictive accuracy of unseen events and comprehensibility at some high cognitive/conceptual level. The quality of the former measure estimates the generalization power of the system; we call it a quantitative property. On the other hand, we call the latter a qualitative property.

The most common experimental methodology is to split the available events into training and testing groups (usually 70% and 30%). Subsequently, the experiment calls for a learning session using the training group, followed by testing using the other group (containing events unseen during the training). Different measures are then used to determine the qualities of a system. For example, for the quantitative properties, Michalski and Chilausky (1980) define a set of conflicting measures under the assumption that in some cases it makes no sense to distinguish between two close diagnoses. However, most researchers use a single measure of accuracy, defined as the ratio of correctly classified events to all testing events: this is the measure we use unless otherwise stated. To measure the qualitative properties, we either list separately the number of rules and conditions used or combine them according to the well-accepted formula shown in section 5.4.

Another important issue is the estimate of statistical measures. We followed those used in the reference publications for each experiment. However, the default approach was to take an average of five independent resampled runs.

7.2. Emerald's robot world

Recently, a report of the AI laboratory at George Mason University (Wnek et al., 1990) evaluated a number of different learning systems using the world of robots from the Emerald system (Kaufman, Michalski, & Schultz, 1989). Since this report provides a detailed description of the experiment, it was relatively easy to repeat exactly the same tests with our system and compare the results directly. In this experiment, we attempted to achieve two goals: to illustrate the system's algorithms and exemplify its comprehensive processing mechanism by tracing one of the experiments; and to gather both qualitative and quantitative results in order to compare them with those published.

This robot world is very suitable for this kind of experiment, since it is moderately complex to allow comparative study, yet simple enough to be illustrated by the diagramatic visualization method. It is described by the following six attributes (we boldface the abbreviations subsequently used in the trace):

Attribute	Values
HeadShape	*Round*, *Square*, *Octagon*
Body	*Round*, *Square*, *Octagon*
Smiling	*Yes*, *No*
Holding	*Sword*, *Balloon*, *Flag*
JacketColor	*Red*, *Yellow*, *Green*, *Blue*
Tie	*Yes*, *No*

The robots were classified into the following five categories created by a human:

Concept	Description
C_1	*Head* is *Round* and *JacketColor* is *Red* or *Head* is *Square* and *Holding* a *Balloon*
C_2	Smiling and *Holding* a *Balloon* or *Head* is *Round*
C_3	*Smiling* and not *Holding* a *Sword*
C_4	*Jacket* is *Red* and no *Tie* or *Head* is *Round* and is *Smiling*
C_5	*Smiling* and *Holding* either a *Balloon* or a *Sword*

The task was to learn a description of each concept while seeing only a varying percentage of the positive and negative examples. There were a total of 432 different robots present in this world. The error rate reported is the average error in recognizing all the 432 (both seen and unseen) events. This measure explicitly estimates the predictive accuracy, while at the same time implicitly judges the generalization and specialization power.

7.2.1. The trace

For the behavior trace we used concept C_1, which can be represented by the following set of rules:

$$[H = R][J = R] ::> C_1,\ [H = S][Ho = B] ::> C_1$$

or, assuming an implicit decision as used in GIL, by the following formula:

$$[H = R][J = R] \vee [H = S][Ho = B]$$

Of the 432 events, 84 satisfy the concept. The training was done using a random 20% of both positive and negative examples: 17 and 70, respectively (see figure 6 for a visualization of the target concept and the training events). The population size was set to 40, initialized equally by both random descriptions and positive training events. The system was set to run 100 iterations. Other implementation parameters were set as follows: $w_1 = w_2 = 0.5$; $w_3 = 0.02$; and the cost was normalized with respect to the highest cost in the current population. The initial application probabilities, along with actual adjusted values (adjustment for the currently average size of the chromosomes; see section 5.4) at the end of this experiment, are presented in table 4. This scaling was computed assuming a desired rate

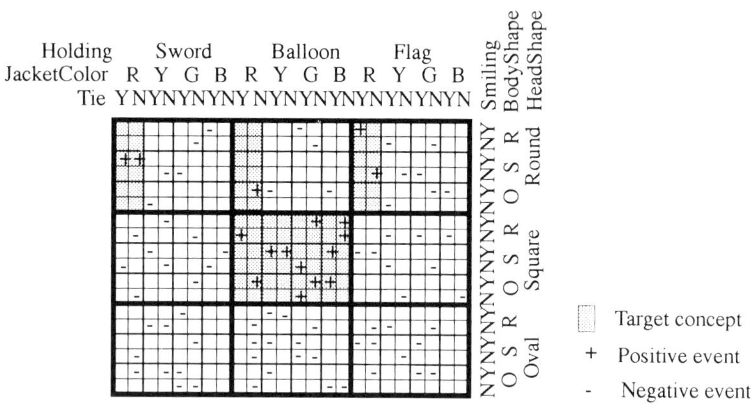

Figure 6. The goal concept and the training events.

Table 4. Application probabilities.

Level	Operator	Initial Values	Final Values
Rule set	Rules exchange	0.20	0.115
	Rules copy	0.10	0.058
	New event	0.40	0.230
	Rules generalization	0.50	0.288
	Rules drop	0.50	0.288
	Rules specialization	0.50	0.288
Rule	Rule split	0.02	0.011
	Condition drop	0.10	0.058
	Turning conj. into disjunc.	0.02	0.011
	Condition introduction	0.10	0.058
	Rule directed split	0.12	0.069
Condition	Reference change	0.02	0.012
	Reference extension	0.03	0.017
	Reference restriction	0.03	0.017

of 80% chromosomes to be updated by the recombination step. The selection probabilities were as follows: 0.2 for a rule selection in "rules exchange"; 0.1 for "splitting a rule" according to two subsets, as opposed to all domain values, for the nominal type, and 0.7 for the linear type; and 0.5 for all probabilities on the condition level.

Before the algorithm could start iterating, it had to perform three initial steps: data compilation, initialization, and initial evaluation. Data compilation involved building the binary coverage vectors for all possible features of the space. Initialization involved setting up the population. Finally, evaluation involved finding the completeness, consistency, and complexity of the rules and accumulating them according to our formula of section 5.4. The best such initial chromosome is illustrated in figure 7, which happened to be a randomly generated description.

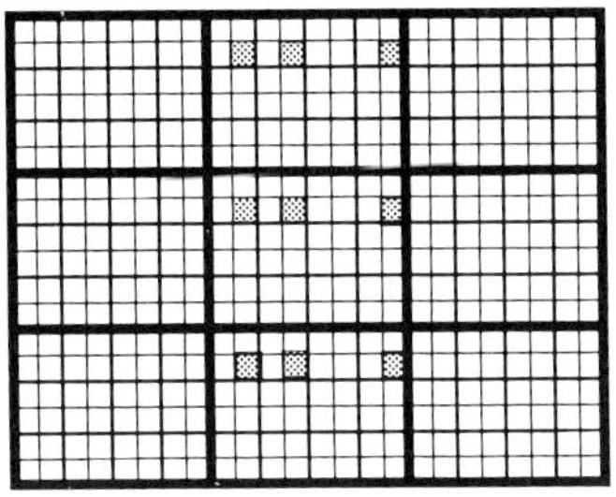

Iteration 0: (from initialization)

Cost = 7 : 1 rule, 5 conditions
Positive coverage = 1
Negative coverage = 0

Rule set:
[B=R][S=N][Ho=B][J=R,Y,B][T=N]

Figure 7. The best initial chromosome.

Each iteration of the genetic algorithm consists of three basic steps: selection, reproduction, and evaluation. However, the special data-compilation method, along with the use of the binary coverage vectors, allows for combining the last two steps as follows. Each operator is followed by a proper update to all the affected vectors and a recalculation of the completeness and consistency measures. Then, the only task of the evaluation step is to calculate the total fitness: there is no pattern matching involved. In the case of an incremental learning (this particular experiment was conducted in the batch mode), every time a new example is presented to the system, all appropriate features (those present in that example) are incrementally updated and propagated to other structures present in the population.

The most important step to explain is reproduction. Normally, reproductive operators are selected based on some static probabilities. However, in our case there are two dynamic factors that affect such probabilities: the rate of chromosome update and the completeness and consistency measures of proper structures. First, the rate of chromosome update (from the previous generation) is compared to that desired (80% in this case, or 32 chromosomes); if it differs by more than some allowable margin, all the application probabilities are accordingly adjusted by a small fraction and the new values replace the old ones. Then each structure of the population is allowed to nondeterministically select operators to be applied to it. Competing operators are those defined for the level of the structure. Moreover, both the generalizing and the specializing operators have their application probabilities adjusted by each structure, before a uniform probability generator decides their actual application. For example, consider a rule set with the following measures: *completeness* = 0.2, *consistency* = 0.9. All operators defined for this level are tried in a random order. Each one actually found applicable updates this complex. The independent operators have probabilities of application exactly as the initial values (possibly adjusted with respect to the desired update rate). The generalizing operators have probabilities of application additionally adjusted by

$$\left(\frac{3}{2} - \textit{completeness}\right) \cdot \left(\frac{1}{2} + \textit{consistency}\right) = 1.82$$

and the specializing operators have the probabilities adjusted by

$$\left(\frac{1}{2} + completeness\right) \cdot \left(\frac{3}{2} - consistency\right) = 0.42$$

In other words, this particular rule set would face an increasing pressure for generalization.

Each operator actually selected for application updates the structure, and then it immediately incrementally updates binary coverage vectors from the structure up to the chromosome level. Also, the appropriate completeness and consistency measures are immediately reevaluated. Such an incremental approach not only reduces the time complexity of evaluations by taking into account some specific information about properties of the operators, but also leaves all the coverages and the measures consistent for the other competing operators. This is very important, since some of them rely on such information for further efficiency improvements. For example, the "directed rule split" operator needs to find a negative event inconsistent with the current complex. If the operator can rely on the coverage vectors, finding such an event is reduced to selecting a random binary one from the negative coverage vector of this complex. Otherwise, the complex would have to be sequentially matched against all possible negative training examples.

Following the initialization and the initial evaluation, the system was run for the specified number of iterations. Not every iteration produced a better chrmosome. There were 11 improvements in 27 iterations, after which the best description exactly matched that of the concept. The first four improvements are illustrated in figure 8.

This problem proved to be extremely simple for our algorithm. After only 27 iterations, or about 3.1 CPU seconds on a DEC3100 station, a complete and consistent description was found. Moreover, this description exactly matched the desired solution—there were no redundancies. This experiment illustrates both the quanitative and qualitative properties of the system.

7.2.2. Comparative experiments

The other systems used in the experiment (reported in Wnek et al., 1990) were rule-based AQ15, neural network BpNet, decision tree with rules generator C4.5, and genetic classifier system CFS. Table 5 reports the average error rate for the five experimental concepts for all five systems while learning one concept at a time (the results of the other four obtained from those published experiments). Surprisingly, GIL produced the highest recognition rate, especially when seeing only a small percentage of the events. This result can be attributed to the simplicity-biased evaluation formula that was used.

Table 6 reports the average acquired knowledge complexity by listing both the average number of rules and the average number of conditions, as learned by all five systems for different learning scenarios in the same experiment. The NR entry indicates that the complexity was large and was not reported in the reference paper. The reason for the higher complexity of the connectionist approach is that this is a non-symbolic system operating on numerical weights rather than on the problem symbols. On the other hand, the high complexity of the CFS approach can be attributed to the fact that the symbolic processing was being done in the representation rather than in the problem space and to the lack of a similar bias for simplicity. This result is rather common for classifier system approaches.

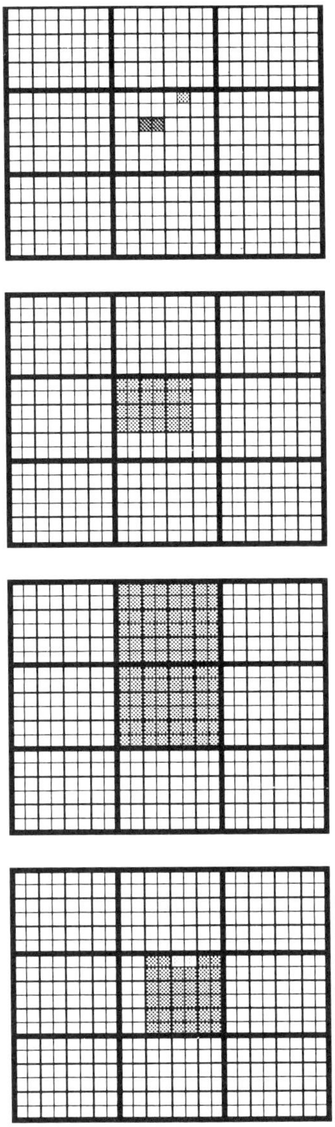

Iteration 1

Cost = 15 : 2 rules, 11 conditions
Positive coverage = 3
Negative coverage = 0

Rule set:
[H=S][B=S][S=Y][Ho=B][J=Y]
∨ [H=S][B=R][S=Y][Ho=B][J=G][T=N]

Iteration 2

Cost = 6 : 1 rule, 4 conditions
Positive coverage = 6
Negative coverage = 0

Rule set:
[H=S][B=S,O][S=Y][Ho=B][J=Y,G,B]

Iteration 3 (an overgeneralization)

Cost = 4 : 1 rule, 2 conditions]
Positive coverage = 12
Negative coverage = 4

Rule set:
[H=R,S][Ho=B]

Iteration 7 (one redundant rule)

Cost = 12 : 2 rules, 8 conditions
Positive coverage = 10
Negative coverage = 0

Rule set:
[H=S][Ho=B][J=Y,G,B]
∨ [H=S][B=R][S=Y][Ho=B][J=G]

Figure 8. The first four improvements.

Therefore, it was a pleasant surprise to find that GIL's knowledge was at the same complexity level as that of AQ15. This demonstrates one of the system's characteristics: the ability to generate easily comprehensible knowledge (measured here by complexity of its VL_1 output). The 1.4/2.6 result in the first column is clearly an oversimplification of the knowledge due to insufficient number of training events (here only about ten negative events were available).

Table 5. Error rate summary in the robot world.

	Learning Scenario (positive%/negative%)				
System	6%/3%	10%/10%	15%/10%	25%/10%	100%/10%
AQ15	22.8%	5.0%	4.8%	1.2%	0.0%
BpNet	9.7%	6.3%	4.7%	7.8%	4.8%
C4.5	9.7%	8.3%	11.3%	2.5%	1.6%
CFS	21.3%	20.3%	21.5%	19.7%	23.0%
GIL	4.3%	1.1%	0.0%	0.0%	0.0%

Table 6. Complexity's summary in the robot world (#rules/#conditions).

	Learning Scenario (positive%/negative%)				
System	6%/3%	25%/10%	50%/10%	75%/10%	100%/10%
AQ15	2.6/4	1.6/3	1.6/3	1.6/3	1.6/3
BpNet	NR	18/29	NR	NR	32/54
C4.5	6.8/12.2	4.4/9.2	4.8/9.2	4.8/9.2	3.8/7.3
CFS	NR	NR	NR	NR	NR
GIL	1.4/2.6	1.6/3	1.6/3	1.6/3	1.6/3

7.3. DNF concepts

Learning DNF descriptions has become a standard way of evaluating different systems. An interesting experiment was reported by Spears and DeJong (1990), in which the authors compared a decision-tree-based ID5R with their own genetic algorithm for supervised concept learning, GABIL (a newer version is described in DeJong and Spears (1991)). The test data for that experiment were a set of random DNF descriptions of a varying complexity. The results represent batch-incremental learning curves: a system's quality measure after seeing n examples is defined as an average recognition of a single unknown random event over the last ten experiments (from $n - 9$ to n). Accordingly, the learning curves are undefined for $n < 10$.

There was a total of six attributes, each having three possible values. Six sets of experiments were conducted, for six randomly constructed DNF concepts xDyC, where $x = 1, 2$ is the number of rules, and $y = 1, 2, 3$ is the number of conditions per rule.

For each experiment, a total of 100 events was chosen randomly. Then, using an increasing number of learning events and just one testing event, the learning curves were constructed using average results over ten independent runs with resampling. For the incremental ID5R, the knowledge was updated incrementally upon a new inconsistent event, while it was generated from scratch in the GABIL system (which does not possess such incremental properties).

We repeated the same experiments in exactly the same environment, using the same batch-incremental mode as GABIL. Because the DNFs actually used were not reported in that paper, we repeated the experiments not only with resampling, but also with randomly

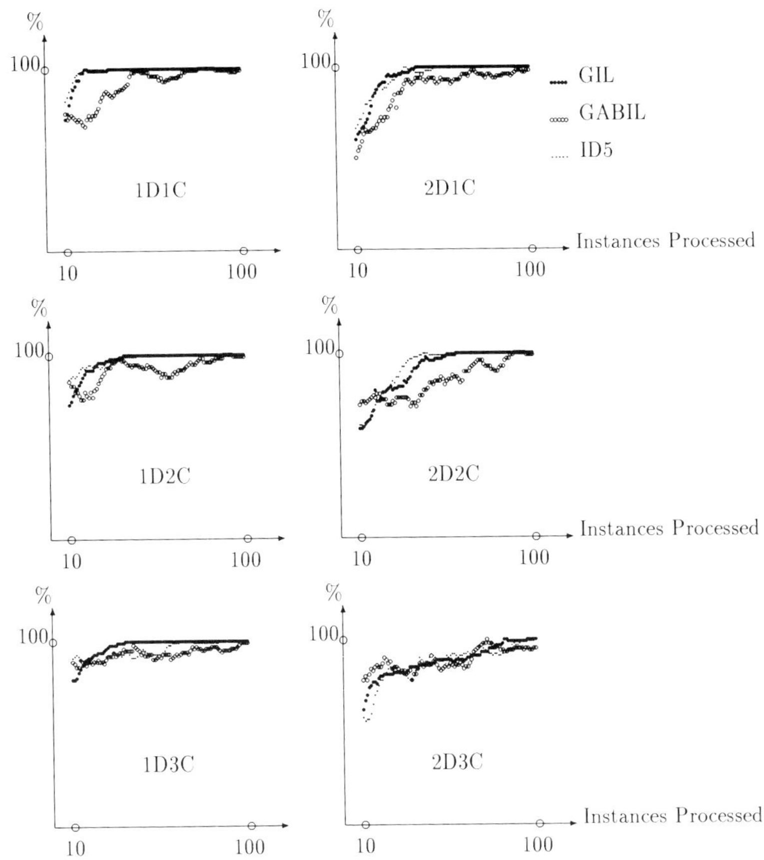

Figure 9. Batch-incremental results on CNF data.

regenerated target descriptions (each run of 100 iterations). Our results, along with the original ones, are presented in figure 9. The original claim of the GABIL system was that it could not learn as well as ID5R on simple concepts, but achieved about the same levels of performance (in some cases, slightly better) as the concepts' complexity increased. Our results show that GIL can do both at the same time: it achieves very high performance for all kinds of problems. It also clearly outperforms GABIL in terms of learning variability. Its smooth learning curve indicates low variability in performance, while the broken curve of GABIL indicates bigger differences from run to run. Moreover, GIL produced descriptions of much lower complexity (not reported for GABIL—private correspondence).

7.4. Multiplexers

The family of multiplexers is another widely used set of data. Each multiplexer is actually a specific case of the more general DNF. For each integer $k = 1, 2, \ldots$ there is a multiplexer

Boolean function defined in the following way: the inputs are the k bits (called addresses), and there are exactly 2^k outputs (called data bits). Accordingly, we have multiplexer f_3 for $k = 1$, f_6 for $k = 2$, f_{11} for $k = 3$, etc. The function of a multiplexer is to activate the data bit whose address (in binary, assuming some ordering of the data bits starting at 0) is specified by the address bits. For example,

$$[A_0 = l][A_1 = 0][A_2 = 1] \vee [A_0 = 0][A_1 = 1][A_3 = 1]$$

$$\vee [A_0 = 1][A_1 = 0][A_4 = 1] \vee [A_0 = 1][A_1 = 1][A_5 = 1]$$

defines the f_6 multiplexer in the VL_1 language, assuming that attributes A_0 and A_1 are the two address bits and A_2 to A_5 are the four data bits.

Many experiments have been documented, mostly using f_6 and f_{11} functions. For example, Koza (1989) describes learning the first of these two, with his LISP-influenced hierarchical genetic approach. He reports a case of learning the actual function after processing 4500 potential solutions, while seeing all 64 (2^6) possible instances. Our own experiments indicate an average learning after seeing only about 2000 individual (40 iterations, database size 50). However, these two results should not be compared directly—Koza apparently used a much larger population size of 300, and the solution was found after 15 iterations. Moreover, there is no indication whether this was the only experiment or the best of a number of experiments. Many classifier systems reported experimentations with a different multiplexer (e.g., Wilson, 1987). However, since CS are in general not full-memory systems, it is again difficult to compare the results directly. Instead, we again decided to illustrate GIL's behavior by tracing its run on f_{11}—this time by observing not the intermediate rule sets but rather the complexity, consistency, and complexity of the best rule set in each generation.

Figure 10 traces a sample run while training with 20% of the available f_{11} events. During this learning session, the exact concept was learned after 1700 iterations. A consistent and complete description of the training events was found shortly after 1000 iterations, and the remaining 700 cycles were required to simplify the generated description. The first of these two graphs traces completeness and consistency of the currently best database individual. The other graph traces the complexity of the best individuals. It is interesting to note that the complexity rises during the learning, as a result of not finding simple enough complete and consistent descriptions, and then decreases—forced down by an increasing cost influence and formation of such better descriptions.

Some quantitative results with f_{11} are reported in table 7. These results are similar to those from other systems (e.g., Quinlan, 1988), but a different experimental methodology does not allow for a direct comparison. They are also similar to reported results of AQ (Wnek & Michalski, 1991), where the authors report 74% accuracy after training with 6% of the events. Some more recent experiments (Janikow, in press) seem to suggest that GIL's results may be greatly improved by adjusting the learning bias. For example, while learning with an increased pressure on consistency, a 100% average accuracy can be achieved for the 20% learning case.

A few interesting differences were observed between f_6 and f_{11} runs. The time necessary for the learning rose from about 10 CPU seconds to about 20 minutes, or from about 100

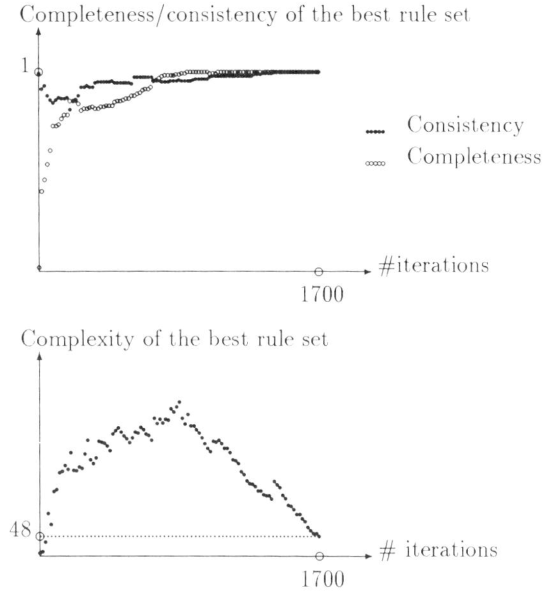

Figure 10. A sample behavior on multiplexer f_{11}.

Table 7. GIL's accuracy on multiplexer f_{11}.

% Training Events	Accuracy
5%	77%
10%	88%
20%	97%

to about 2000 iterations. This increase can be attributed to the increased event space (64 vs. 2048) and the increased search space ($\sim 10^{200}$ vs. $10^{60,000}$). More directly, this caused an increased number of iterations and a noticeable increase in the average complexity of chromosomes. A further investigation is needed to determine the scaling characteristics of the system.

Yet another interesting difference between the two multiplexers can be observed by comparing the accuracy while learning with a small percentage of the available events: f_{11} achieves high rates much more quickly. The reason for such a behavior seems to be the ratio of the concept complexity to the size of event space, which is about 0.313 for f_6 and 0.023 for f_{11}. This raises an interesting hypothesis about estimating the difficulty of generating descriptions: the difficulty is proportional, or at least highly correlated, to the ratio of the concept's complexity and the size of the event space. We hope to further investigate this assertion in the figure.

Table 8. Summary of the breast cancer experiment.

System/Method	Complexity	Accuracy
Human experts	Not reported	64%
AQ15/full rule set	41 rules/160 conditions	66%
AQ15/best rule only	2 rules/7 conditions	68%
Assistant/without tree pruning	63 leaves/120 nodes	67%
Assistant/with tree pruning	9 leaves/16 nodes	72%
GIL	36 rules/128 conditions	65%
GIL/with emphasized cost	10 rules/27 conditions	67%

7.5. Breast cancer

One of the most popular natural domains used in experiments with inductive learning systems is the breast cancer data. It contains 286 descriptions of female patients, classified as either developing or not developing a recurrence of breast cancer after a five-year period following the first surgery. The descriptions are generated using nine attributes, with an average of 5.8 values per domain. This descriptive language was found to be inconsistent, meaning that some patients having exactly the same description were classified differently. Such a situation puts an extra burden on the learning system.

An excellent paper by Michalski and colleagues (1986) lists both quantitative and qualitative results on this data set, while using the AQ15 system with a rule truncation mechanism and a decision tree system ASSISTANT (this version generates low-complexity binary trees and employs a tree truncation technique). In addition, this paper also reports the accuracy of human experts. We repeated these experiments (for 1000 iterations) and report all such results in table 8. Our second run was performed with an increased cost influence on the evaluation (increased w_3 parameter). Both results indicate the applicability of our approach in the case of natural domains as well. For compatability with the complexity results reported for the other systems, we ran GIL twice, each time learning one of the two possible concepts and summarizing the #conditions and #rules.

7.6. Incremental learning

Incremental learning capabilities are important attributes of a learning system. We expected our approach to possess such properties naturally, since it is based on generalization and specialization of the current hypotheses and does not explicitly favor either of these two classes of actions at any time. To evaluate this assertion, we performed both batch and incremental experiments to produce the learning curves of f_6. Both learnings were performed at 5% increments of the available training events: in the batch mode, each new experiment started with newly generated population; and in the incremental mode, each new experiment started with the population generated at the end of the previous learning session with fewer training events. All experiments were repeated five times with resampling, with population size set to 40 and run for 100 interations. Figure 11 presents these curves. It is interesting to note that there was no significant difference in the quantitative

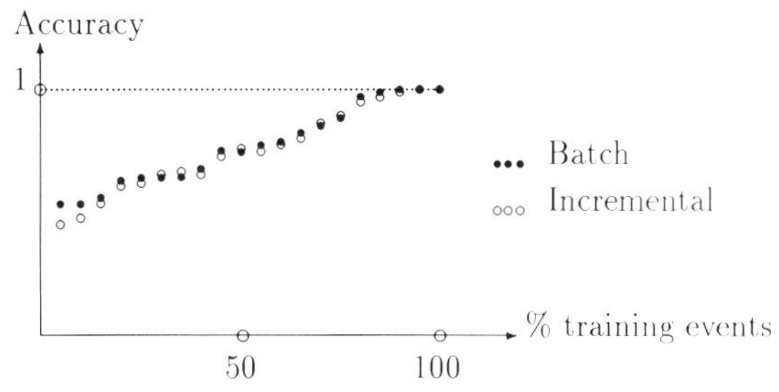

Figure 11. Comparison of the batch and incremental learning.

performance, suggesting GIL can be used in incremental environments. Moreover, observed faster convergence in the incremental mode suggests that the reused knowledge in forms of the initial population was processed very usefully, indicating the system's ability to process initial hypotheses and maintain knowledge in a dynamic environment. Again, more experimentation is needed to further investigate these properties.

8. Conclusions and further research

We have described a novel approach to full-memory, supervised inductive learning in attribute-based spaces that uses a knowledge-intensive genetic algorithm. Such an algorithm follows the ideas of traditional GAs, but replaces the domain-independent search by a search specific to the inductive learning methodology. This approach represents an abstraction of the traditional genetic algorithm to the symbolic level. Initial results show that GAs can be successfully applied to more complex, non-numerical tasks by defining the algorithm at the conceptual level of the problem. This allows for processing high-level structures using the problem-specific methodology and rich heuristics. Moreover, such an abstract view provides for the same clear separation of different systems' components as found in AI production systems. This modularity, in turn, allows for transparent applications of similar designs to other domains. Some of the most interesting properties of this approach are the clearly understood processing mechanisms, defined on the same language as that of input/output, and the low complexity of the generated knowledge. This complexity, measured here by number of rules and conditions, contributes to the higher comprehensibility of the output.

When pursuing this challenge, we did not attempt to produce a system that could compete with the existing symbolic systems (AQ and ID based), especially in terms of learning time complexity. Our goal was rather to investigate the potential of such a method of abstracting genetic algorithms, which may be carried to other domains. Nevetheless, by designing efficient data-compilation methods aimed at reducing the system's complexity, and by using more "intelligent" operators than those in the traditional GAs, we were able to tackle a number of interesting problems in a reasonable time. Moreover, since genetic algorithms are naturally suited for parallel architectures, we may hope that such reimplementations, along with new technological advances, may be faster without any additional efforts.

The system also shows significant potential from the machine learning point of view. First, it does not assume attribute independence, as the ID-based systems do. Second, it provides for a better match between problem-specific heuristics and actions taken than the AQ-based systems do. Finally, it extends the AQ's ideas of exploring a number of simultaneous directions to a more powerful platform, allowing for both competition and cooperation (by information exchange).

The current complexity, as well as the overall quality, seems to be unstable and to vary among different problems, different parameters used, and even different runs of the same experiment. Making the algorithm more stable requires an extensive study of its characteristics. Other important issues to be addressed include learning multiple concepts and dealing with noisy information. These issues are further discussed below along with some proposed solutions. Yet other, more challenging, future research involves using more powerful languages than VL_1 and adapting the system in a non-full-memory environment.

One should point out that most of the testing data were not suited to explore the full potential of this system: most experiments were conducted with two- or three-valued domains. Such domains are much more suitable to the other learning systems, while this approach can fully explore spaces with larger, typed domains. Actually, among the other known learning systems, only AQ and SAMUEL try to accommodate knowledge of domain types, which can be quite valuable when the domains grow. However, AQ does so only after learning the initial complete and consistent descriptions (a recent development in AQ systems attempts to use this information in a second stage of its two-tiered learning).

8.1. Multiple domains

The current system assumes single concepts only, and proceeds by learning only rules associated with the single decision. This lets us treat the decisions implicitly and then simplify the definitions and the implementation by processing VL_1 complexes in place of rules. For the sake of simplicity and continuity, we will try to preserve this property while generalizing the approach to learning multiple decisions. If we assume independence of different class descriptions, we may apply subsequent learning sessions of the same algorithm, one session per class. Then, during a learning sesson for class n, the examples of category n are considered as positive events, while the examples of all other categories are considered as the current negative events. The same idea may be used differently: we may conduct the learning sessions simultaneously in different populations, with one population assigned to one class being learned.

When discussing such generalizations, we must consider a related issue: treating descriptions that are incomplete and inconsistent with respect to the problem space. As we mentioned before, a rule-based framework is not well suited for learning descriptions with those properties, which are often relaxed. However, in such a case, a previously unknown event may be recognized by none or multiple descriptions. Then, a special arbitration protocol must be employed. The simplest such protocol simply returns such events as unrecognizable. This improves correct recognition rate, but also increases overall indecision. Moreover, this approach can treat descriptions of different classes independently. A more sophisticated protocol employs some flexible (e.g., probabilistic) measures, and thus changes the conceptualization view to non-crisp. However, this also suggests that the descriptions should not be generated independently, but in a common context. Then each rule set must be evaluated in the context of the descriptions for the other classes. This suggests learning all descriptions in a single population, with a chromosome being a set of descriptions now. We can still preserve the implicitness of decisions if we introduce an additional syntactic level: the decision set set level. Then, while learning descriptions of n categories, each chromosome becomes a set of descriptions, with each description associated with exactly one of the categories. In other words, a chromosome becomes a set of the previously defined chromosomes. Then, all operators that were defined as acting on two chromosomes will act on two rules sets associated with the same decision. All other operators stay exactly as defined previously. However, we need new operators to act on the new level. The only proposed operator has an independent character, and it exchanges one or more whole-category descriptions between two chromosomes. In other words, this operator works at the level of granularity of a whole rule set associated with one implicit decision.

As to the non-crisp arbitration protocol, a very nice solution is used in the AQ family—the two-tiered conceptualization view. We suggest using the same approach in such an extended architecture, which could provide an additional advantage. The two-tiered view could be used during the learning process, while it is used only in a follow-up step in the AQ system. Because a given problem specifies a priori the number of categories to be conceptualized, we can easily extend the chromosome implementation of section 6.2 to a vector of such, where a vector position is associated with the decision number.

8.2. *Other issues*

One of the major disadvantages of the current implementation is its high parameterization: there are about 40 input parameters that must be specified, with most of them being continuous probability values. Under such conditions, it seems highly unlikely that the proper combination for a given run will be selected. This is the reason for the poorer performance noticed on more complex problems (e.g., multiplexer f_{11}). Moreover, on some occasions we observed an improved performance after slightly changing the initial probabilities. However, a much more extensive and systematic experimentation is necessary to determine the actual source of such variations: randomness of the runs or some problem-specific characteristics.

To deal with the problem of the size of the parameter space, it is necessary to explore inter- and intra-dependencies between such parameters. For example, all rule-set level application probabilities should be specified with respect to each other and, as a group, with respect to those of other groups. The dependence of the selection probabilities on the problem size should be explored. However, establishing such relations requires an extensive testing over a wide variety of different problems. To deal with the second problem (dependency on characteristics of the problem), the choice of such abstracted parameters should be further associated (in addition to the dynamic method of section 5.5) with the problem complexity (determined dynamically in the process of learning) and some learning criteria (specified by the user or some other requirements). Then all these parameters could be replaced by few conceptual ones, for example, desired type of descriptions (such as specific vs. general, or low attribute cost vs. low descriptive cost). Such an abstraction would provide for both easier use and more efficient performance. To deal with other non-probabilistic parameters, most of which control the learning bias, it is necessary to study such bias extensively. Some recent results show that adjusting the bias appropriately can improve both the speed and the quality of learning (Janikow, in press).

While it might be conceptually advantageous to assume the existence of noise-free data, it is often unrealistic under natural conditions. Therefore, an artificial system aimed at working in a natural domain should deal with these issues. Some researchers addressed the problem of noisy features. For example, Clark and Niblett (1987) discuss the effect of noise on induction and present their probabilistic way of dealing with the problem. Quinlan (1990), showed how to deal with the effect of noisy data in decision trees. The two-tiered representation of AQ family of systems applies to noise effect reduction as well. Looking at these attempts, it is fair to say that noise accommodation generally employs some kind of numerical approach. Our simplified approach assumes a crisp concept representation with rule-based conceptualization and is not well suited for dealing with noise: the most appealing approach would be to combine rule-coverage information with rule complexity in such a way that rules with very low positive coverage become more costly and more likely to be removed. The same strategy applies also to all of the proposed generalizations to deal with multiple concepts. However, in the case of the extended chromosome architecture, the two-tiered description may itself be more valuable as a noise-accommodation agent, since this approach was shown to improve noisy recognition in the AQ15 system.

We mentioned in section 5.5 that pursuing atomicity and efficiency over complexity made us neglect the inductive resolution rule. It would be an interesting study to compare the behavior of the current system with another one implementing this operator, along with other possible new operators. Such a study should concentrate on both the quantitative and qualitative properties, as well as possible quality vs. time trade-off.

In addition, some of the rule set level operators described in section 5.5 could be differently defined, depending on the selection of applicable rules. For example, the "rules exchange" operator could overlook the selection probability and always select a single rule. On the other hand, the "rules copy," "rules generalization," "rules drop," and "rules specialization" could be enhanced by such probabilities, instead of always choosing a fixed number of rules (one or two, depending on the nature of the operator). Again, an extensive study is required to establish values of such new operators in relation to those presently used.

The ideas used here can be seen as a very specific case of more general ideas presented by Davis (1991), where he calls for exploring combinations of the traditional domain-independent genetic algorithms with some known problem-specific methods (the hybrid approach). The only difference between this system and those ideas lies in the initialization. Davis argues that the first population should be filled by chromosomes representing properly represented results of some known fast systems. The argument for such an enhancement is that then the genetic algorithm can only improve such ad hoc solutions. Therefore, it is guaranteed to perform at least as well as such other systems. In our domain, there are such programs. For example, most decision tree systems are very efficient and produce assertions easily convertible to a rule-based format. Since our initial experiments indicate the system's ability to process some initial hypotheses, there is a potential for a significant performance and time improvement. Moreover, another interesting idea is to use decision-tree-based operators, which partition the current event subspace, instead of the less sophisticated "rule split" operator.

Acknowledgments

The author wishes to thank all of the reviewers for their valuable comments. This work was partially supported by a computational grant from the Microelectronic Center of North Carolina.

References

Antonisse, H.J., & Keller, K.S. (1987). Genetic operators for high level knowledge representation. *Proceedings of the Second International Conference on Genetic Algorithms* (pp. 69–76). Hillsdale, NJ: Lawrence Erlbaum Associates.

Baker, J.E. (1987). Reducing bias and inefficiency in the selection algorithm. *Proceedings of the Second International Conference on Genetic Algorithms* (pp. 14–21). Hillsdale, NJ: Lawrence Erlbaum Associates.

Booker, L.B. (1989). Triggered rule discovery in classifier system. *Proceedings of the Third International Conference on Genetic Algorithms* (pp. 265–174). San Mateo, CA: Morgan Kaufmann.

Clark, P., & Niblett, T. (1987). Induction in noisy domains. In I. Bratko & N. Lavrac (Eds.), *Progress in machine learning*. Sigma Press.

Davis, L. (Ed). (1991). *Handbook of genetic algorithms*. New York: Van Nostrand Reinhold.

DeJong, K.A. (1985). Genetic algorithms: A 10 year perspective. *Proceedings of the First International Conference on Genetic Algorithms* (pp. 169–177). Hillsdale, NJ: Lawrence Erlbaum Associates.

DeJong, K.A. (1988). Learning with genetic algorithm: An overview. *Machine Learning, 3 (2/3)*, 121–138.

DeJong, K.A. (1990). Genetic-algorithm-based learning. In Y. Kondratoff & R.S. Michalski (Eds.), *Machine learning: An artificial intelligence approach* (Vol. 3). San Mateo, CA: Morgan Kaufmann.

DeJong, K.A., & Spears, W.M. (1991). Learning concept classification rules using genetic algorithms. *Proceedings of the International Joint Conference on Artificial Intelligence* (pp. 651–656). San Mateo, CA: Morgan Kaufmann.

Forrest, S. (1985). Implementing semantic networks structures using the classifier system. *Proceedings of the First International Conference on Genetic Algorithms* (pp. 24–44). Hillsdale, NJ: Lawrence Erlbaum Associates.

Goldberg, D.E. (1985). Genetic algorithm and rule learning in dynamic system control. *Proceedings of the First International Conference on Genetic Algorithms* (pp. 8–15). Hillsdale, NJ: Lawrence Erlbaum Associates.

Goldberg, D.E. (1989). *Genetic algorithms in search, optimization & machine learning*. Reading, MA: Addison-Wesley.

Goldberg, D.E., & Holland, J.H. (1988). Genetic algorithms and machine learning. *Machine Learning 3 (2/3)*, 95–99.

Grefenstette, J. (1987). Incorporating problem specific knowledge into genetic algorithms. In L. Davis (Ed.), *Genetic algorithms and simulated annealing*. London: Pitman.

Grefenstette, J. (1991). Lamarckian learning in multi-agent environments. *Proceedings of the Fourth International Conference on Genetic Algorithms* (pp. 303–310). San Mateo, CA: Morgan Kaufmann.

Green, D.P., & Smith, S.F. (1985). A genetic system for learning models of consumer choice. *Proceedings of the Second International Conference on Genetic Algorithms* (pp. 217–223). Hillsdale, NJ: Lawrence Erlbaum Associates.

Holland, J.H. (1975). *Adaptation in natural and artificial systems*. Ann Arbor, MI: University of Michigan Press.

Holland, J.H. (1986). Escaping brittleness. In R.S. Michalski, J. Carbonell & T. Mitchel (Eds.), *Machine learning: An artificial intelligence approach* (Vol. 2). San Mateo, CA: Morgan Kaufmann.

Holland, J.H., Holyoak, K.J., Nisbett, R.E., & Thagard, P.R. (1986). *Induction*. Cambridge, MA: MIT Press.

Janikow, C.Z. (1991). *Inductive learning from attribute-based examples: A knowledge-intensive genetic algorithm approach*. Doctoral dissertation, University of North Carolina at Chapel Hill, Department of Computer Science.

Janikow, C.Z. (in press). Some experiments with a stochastic production system for supervised inductive learning. In *Artificial intelligence: Methodology, systems, applications* (Vol. 5). Amsterdam: North-Holland

Kaufman, K.A., Michalski, R.S., & Schultz, A.C. (1989). *EMERALD 1: An integrated system of machine learning and discovery programs for education and research.* (Technical Report MLI-89-12). Fairfax, VA: Center for AI, George Mason University.

Koza, J.R. (1989). Hierarchical genetic algorithms operating on populations of computer programs. *Proceedings of the Eleventh International Joint Conference on Artificial Intelligence* (pp. 768–774). San Mateo, CA: Morgan Kaufmann.

Liepins, G.E., & Wang, L.A. (1991). Classifier system learning of Boolean concepts. *Proceedings of the Fourth International Conference on Genetic Algorithms* (pp. 318–323). San Mateo, CA: Morgan Kaufmann.

Michalewicz, Z., & Janikow, C.Z. (1991). Genetic algorithms for numerical optimization. *Statistics and Computing, 1 (1)*, 75–89.

Michalski, R.S. (1983). Theory and methodology of inductive learning. In R.S. Michalski, J. Carbonell, & T. Mitchel (Eds.), *Machine learning: An artificial intelligence approach* (Vol. 1). San Mateo, CA: Morgan Kaufmann.

Michalski, R.S. (1986). Understanding the nature of learning. In R.S. Michalski, J. Carbonell, & T. Mitchell (Eds.), *Machine learning: An artificial intelligence approach* (Vol. 2). San Mateo, CA: Morgan Kaufmann.

Michalski, R.S., & Chilausky, R.L. (1980). Learning by being told and learning from examples: An experimental comparison of the two methods of knowledge acquisition in the context of developing an expert system for soybean disease diagnosis. *International Journal of Policy Analysis and Information Systems, 4 (2)*, 125–161.

Michalski, R.S., Mozetic, I., Hong, J., & Lavrac, N. (1986). *The AQ15 inductive learning system: An overview and experiments* (Technical Report UIUCDCS-R-86-1260). Urbana, IL: Department of Computer Science, University of Illinois at Urbana-Champaign.

Quinlan, J.R. (1986). Induction of decision trees. *Machine Learning, 1 (1)*, 81–106.

Quinlan, J.R. (1988). An empiricial comparison of genetic and decision-tree classifiers. *Proceedings of the Fifth International Conference on Machine Learning* (pp. 135–141). San Mateo, CA: Morgan Kaufmann.

Quinlan, J.R. (1990). Probabilistic decision trees. In Y. Kondratoff & R.S. Michalski (Eds.), *Machine learning: An artificial intelligence approach (Vol. 3)*. San Mateo, CA: Morgan Kaufmann.

Rendell, L.A. (1985). Genetic plans and the probabilistic learning system: Synthesis and results. *Proceedings of the First International Conference on Genetic Algorithms* (pp. 60–73). Hillsdale, NJ: Lawrence Erlbaum Associates.

Rendell, L., Cho, H., & Seshu, R. (1989). Improving the design of similarity-based rule-learning systems. *International Journal of Expert Systems, 2 (1)*, 97–133.

Schaffer, J.D. (1985). Learning multiclass pattern discrimination. *Proceedings of the First International Conference of Genetic Algorithms* (pp. 74–79). Hillsdale, NJ: Lawrence Erlbaum Associates.

Sedbrook, T.A., Wright, H., & Wright, R. (1991). Application of genetic classifier for patient triage. *Proceedings of the Fourth International Conference on Genetic Algorithms* (pp. 334–338). San Mateo, CA: Morgan Kaufmann.

Smith, S.F. (1980). *A learning system based on genetic algorithms.* Doctoral dissertation, University of Pittsburgh, Department of Computer Science.

Spears, W.M., & DeJong, K.A. (1990). Using genetic algorithms for supervised concept learning. *Proceedings of the Second International Conference on Tools for AI* (pp. 335–341). Washington, DC: IEEE Computer Society Press.

Weiss, S.M., & Kapouleas, I. (1989). An empirical comparison of pattern recognition, neural nets, and machine learning classification methods. *Proceedings of the Eleventh International Joint Conference on Artificial Intelligence* (pp. 781–787). San Mateo, CA: Morgan Kaufmann.

Weiss, S.M., & Kulikowski, C.A. (1990). *Computer systems that learn.* San Mateo, CA: Morgan Kaufmann.

Wilson, S. (1987). Classifier systems and the animat problem. *Machine Learning, 2 (3)*, 199–228.

Wnek, J., & Michalski, R. (1991). *Hypothesis-driven constructive induction in AQ17: A method and experiments* (Technical Report MLI-91-4). Fairfax, VA: Center for AI, George Mason University.

Wnek, J., Sarma, J., Wahab, A., & Michalski, R.S. (1990). Comparing learning paradigms via diagramatic visualization. In M. Emrich, Z. Ras, & M. Zemankowa (Eds.), *Methodologies for intelligent systems 5.* Amsterdam: North-Holland.

Received October 7, 1991
Accepted March 27, 1992
Final Manuscript July 23, 1992

Machine Learning, 13, 229-257 (1993)

Competition-Based Induction of Decision Models from Examples

DAVID PERRY GREENE dglv@andrew.cmu.edu
STEPHEN F. SMITH sfs@isl1.ri.cmu.edu
The Robotics Institute, School of Computer Science, Carnegie Mellon University, Pittsburgh, PA 15213

Abstract. Symbolic induction is a promising approach to constructing decision models by extracting regularities from a data set of examples. The predominant type of model is a classification rule (or set of rules) that maps a set of relevant environmental features into specific categories or values. Classifying loan risk based on borrower profiles, consumer choice from purchase data, or supply levels based on operating conditions are all examples of this type of model-building task. Although current inductive approaches, such as ID3 and CN2, perform well on certain problems, their potential is limited by the incremental nature of their search. Genetic algorithms (GA) have shown great promise on complex search domains, and hence suggest a means for overcoming these limitations. However, effective use of genetic search in this context requires a framework that promotes the fundamental model-building objectives of predictive accuracy and model simplicity. In this article we describe COGIN, a GA-based inductive system that exploits the conventions of induction from examples to provide this framework. The novelty of COGIN lies in its use of training set coverage to simultaneously promote competition in various classification niches within the model and constrain overall model complexity. Experimental comparisons with NewID and CN2 provide evidence of the effectiveness of the COGIN framework and the viability of the GA approach.

Keywords. Genetic algorithms, symbolic induction, concept learning

1. Introduction

Many practical decision-making problems involve prediction in complex, ill-understood domains where the principal source of predictive knowledge is a set of examples of observed, prior-domain behaviors. These examples typically relate the values of a set of measured situational features (or variables) to specific outcomes, and the critical prerequisite for decision-making is construction of a model that captures the regularities in these data. Estimation of loan risk based on borrower profiles, prediction of consumer preferences given past purchase data, and forecasting resource supply levels based on operating conditions are all examples of this type of problem. We can identify two desirable properties of a decision model in such contexts: (1) to accurately predict behavior in new situations and (2) to provide descriptive insight into the underlying dynamics of the problem domain by fitting the most concise model given the data.

The task of constructing useful decision models from examples in realistic environments is complicated by problem complexity. Sources of problem complexity include:

- Generating function. Often the underlying behavior comes from an unknown function that is nonlinear and that may have interacting, nonhomogeneous variables.[1]

- Noise. Further complicating the model-generation problem is the possibility of noise in the form of errors in recording the attribute values, executing the example-generating behavior, or classifying the example.
- Scale. The data set may have many examples measured along a large number of dimensions with mixed data types. This gives rise to a combinatorially large space of possible models that must be searched.

The goal of constructing decision models from examples has been approached from different perspectives. Statistical methods (e.g., regression techniques) have proved effective in generating accurate predictive models that are robust in the presence of noise. However, their effectiveness is limited to domains where underlying modeling assumptions (e.g., linear feature interactions) are reasonably satisfied, and even there, statistical models provide little descriptive insight. Alternatively, research in concept learning has emphasized mechanisms for generating symbolic decision models that map classes of situations to specific categories or values. A symbolic representation improves descriptive potential, but it also increases the combinatorics of the model-building problem. Problem size, coupled with noise and non-linear feature interactions, can create problems for many current symbolic induction approaches. Much effort has focused (often successfully) on augmenting the basic search with post-processing and other refinement techniques.

We argue that much of the difficulty in dealing with complexity stems from the bias inherent in the deterministic, stepwise search techniques exploited by these approaches. In contrast to these approaches, genetic algorithms (GAs) have shown great promise in complex search domains. A GA operates in an iterative improvement fashion where search is probabilistically concentrated toward regions of the model representation space that have been found to produce good classification behavior. The properties of genetic search match up well with the above identified characteristics of practical classification problems. However, effective use of genetic search in this context requires a framework that promotes the fundamental model-building objectives of predictive accuracy and model simplicity.

In this article we describe COGIN, a GA-based inductive system that provides such a framework. The design of COGIN is tailored to exploit the special character of problems that require model induction from a set of pre-specified training examples. The result is a fairly significant departure from standard GA-based learning paradigms. The fundamental organizing principle of COGIN is the use of training set coverage as both an explicit constraint on model complexity (size) and a basis for creating a pressure toward appropriate diversity within the model. Survival of a given individual is thus a direct function of its ability to fill a particular classification niche in the evolving model. We present comparative experimental results conducted with COGIN and two other well-known symbolic induction systems that provide evidence of the utility of this approach. More specifically, we demonstrate the leverage provided by genetic search across model-building problems of varying complexity and show performance differential trends that suggest that this leverage can increase with problem complexity.

The remainder of this article is organized as follows. In section 2, we define the symbolic induction problem to be addressed, describe the dominant current approaches to this problem, and identify their potential weaknesses. Section 3 provides background concerning our previous work in applying GAs to symbolic induction problems, tracing the evolutionary progression to the current COGIN system. Section 4 presents the principal components of COGIN and discusses its relationship to other work in GA-based learning. In section 5, we present the results of initial experimental comparisons of COGIN, NewID, and CN2 across both a range of artificial test problems and an externally provided data set representative of an actual resource supply forecasting application. We conclude in section 6 with a brief discussion of outstanding research issues.

2. Induction of symbolic decision models from examples

2.1. *The structure of the problem*

In defining the symbolic induction problem of interest, we first assume the existence of a pre-classified data set. "Pre-classified" normally refers to a set of examples in which each example has been assigned a discrete class as the criterion variable. However, this may also refer to the more general case of a data set in which there is a specified dependent variable. This distinguishes the problem of constructing models from a pre-classified data set from the cluster problem, where no dependent variable is identified. More precisely, a pre-classified data set consists of a collection of individual examples or observations where each example is described by a set of features (also referred to as predictor or independent variables) and an associated classification value (also referred to as the criterion or dependent variable). Typically there will be multiple predictor variables of mixed data types and a single criterion variable. The criterion variable is often binary or discrete-valued, but might be continuous. For instance, consider the data set of business loan profiles depicted in figure 1.

An example extracted from this data base might be specified as follows:

industry = electronics, sales = 15M, growth = 15%, debt/equity = 0.6 → loan_risk = low

In this example the predictor variables are industry, sales, growth, and debt/equity ratio. The criterion variable would be "loan-risk," with each example classed as either "high," "medium," or "low" risk—possibly based on the previous judgment by an expert or an actual loan default.

Given a pre-classified data set, the induction task is one of finding a set of feature patterns that partition the elements of the data set into the desired classes and provides a basis for effective classification of future unseen examples expressed in terms of the same predictor

rec_num	industry	sales($K)	growth	debt/equity	loan_risk
...					
0032	electronics	15,000	0.15	0.6	low
0033	oil&gas	137,000	0.09	1.1	medium
0034	restaurant	3,500	0.12	0.45	medium

Figure 1. Data set of business loan profiles.

variables. Such a decision model commonly takes the form of a set of IF → THEN rules; for example,

IF [industry=oil_and_gas] and [debt/equity < 1.2] and [cash_flow=high]
THEN loan_risk ⇒ low
IF [industry=technology] and [debt/equity > 2] and [company_age < 5 yrs]
    and [current_growth < 20%]
THEN loan_risk ⇒ high

In this case, each rule consists of a conjunction of specific attribute-value conditions and an associated outcome, and different rules with the same associated outcome are used to express disjunctive sets of classifying conditions. An ordering assumption might also be imposed over the rule set, in which case the specification of condition/outcome pairs depends additionally on an implicit "IF → Then, else" structure, (e.g., if ⟨condition a⟩ then class-x, else if ⟨condition b⟩ ....).

As indicated at the outset of this article, we can identify two dimensions along with the utility, or quality, of a given decision model can be measured: (1) the model's predictive accuracy and (2) the descriptive insight that the model provides. During model construction, or training, predictive ability can be estimated by how accurately the decision model classifies the available set of training examples. Ultimately, the performance of the system will be based on the predictive accuracy of the final model when applied to a holdout sample of examples not seen during training. Insight is difficult to operationalize, but a common surrogate is to use "simplicity"—that is, prefer smaller models, or, in the case of a rule-based model representation, prefer models with fewer rules.

With respect to the size of a model, one additional distinction can be made regarding "ordered" versus "unordered" rule sets. Ordered rule sets are based (and constructed) on an explicit assumption of sequential interpretation (i.e., the rule set has an implicit If-Then-Else structure). Within unordered rule sets, alternatively, each rule fully specifies the circumstances under which it should be applied, and an ordering is imposed only as needed to resolve situations of conflicting matches. Because of the explicit dependence on context in specifying rules, ordered models will require fewer rules than unordered models. However, this additional compactness does not necessarily translate to increased descriptive insight or model comprehensibility. It is difficult to evaluate an individual rule of an ordered model out of context of the rules that precede it, while in the case of an unordered model the rules themselves have been validated independently.

2.2. *Prototypical approaches*

For the problem of learning decision models from examples under realistic conditions, the most successful and widely applied approaches are those which incrementally generate decision trees and decision lists (MacDonald & Witten, 1989). The prototypical examples of these approaches are ID3 (Quinlan, 1984; Quinlan, 1986) and CN2 (Clark & Niblett, 1989), respectively. Under noise-free conditions, these systems attempt to induce a generalized description that is both complete, by including all positive examples of the concept, and consistent, by excluding all negative examples. For noisy environments, CN2 relaxes the consistency requirement, while the C4 varient of the ID3 relies on tree pruning and other post-processing techniques.

Both ID3 and CN2 conduct a stepwise search for a solution in their respective problem spaces. In the case of ID3, a decision tree is developed by first identifying the individual feature value condition that maximally separate the positive and negative examples, and then recursively finding the next best split for each branch that leaves one or more examples unclassified. CN2 (an extension of Michalski's AQ algorithm (Michalski & Chilauski, (1980)) alternatively constructs a decision list model, represented as a sequence of rules (called complexes) that denote alternative conjunctive conditions of feature values. CN2 has an *ordered* and *unordered* variant, each of which uses a pruned, general-to-specific beam search to construct complexes incrementally. In the ordered case, the training examples matched by a generated complex are removed from further consideration; in the unordered case, all examples are considered during generation of each complex.

Reliance on incremental search methods enables systems like ID3 and CN2 to effectively manage the combinatorics of the model inference problem. At the same time, the utility of the decision models generated using these methods depends on the nature of the modeling problem. Stepwise search methods assume that the problem features are independent, additive, and non-epistatic[2] in their impact on decisions. In more complex problem domains, where these assumptions are not met (e.g., noisy data, feature interactions), the best path at each step can lead the search astray. The search procedure is locally optimal, but it is not necessarily globally optimal.

3. Induction of decision models using genetic algorithms

In contrast to the deterministic, stepwise search mechanisms described above, the GA is a probabilistic, competition-based search technique based on the concept of adaptive efficiency in natural organisms (Holland, 1975). A GA maintains a collection of candidate solution structures (the population) and proceeds by repeatedly selecting structures (individuals) from the current population according to relative performance (fitness) and applying idealized reproductive operators (principally recombination) to produce new structures (offspring) for insertion into the population and testing. GAs gain their power from an ability to implicitly exploit performance information about the exponentially large number of solution "building blocks" (e.g., rules, rule clauses) present in the structures composing the current population in order to focus the generation of new candidate solutions. Specific building blocks are propagated through the population over time in direct proportion to the observed performance of the structures containing them. From a sampling perspective, this implicit parallelism leads to increasing concentration of trials to those regions of the search space observed to be most profitable. The potential of genetic search has been demonstrated in a variety of complex learning contexts (Booker, 1982; Goldberg, 1985; Grefenstette, 1990; Smith, 1983; Wilson, 1987). By their iterative and probabilistic nature, GAs are relatively insensitive to both noise and the manner in which generalizing data are presented. The reader is referred to Goldberg (1989) for an introduction to the theory and practice of GAs.

3.1. Basic choices

Approaches to GA-based inductive learning can be categorized along two dimensions, corresponding to two principal design decisions that must be taken. The first concerns the representation of the decision model to be learned. Given the simple syntactic character of the basic recombination operator (i.e., crossover), its straightforward application to complex model representations is problematic from the standpoint of consistently generating new rule structures that are semantically meaningful (which is fundamental to furthering the search). Historically and still predominantly, this representation problem has been solved through the use of simple, fixed-length condition/action rule formats defined with respect to an ordered, binary encoding of situational features and possible actions. Such rule-based model representations permit direct application of standard crossover operators with assurance that all generated offspring will be viable rule structures. Some more recent efforts (e.g., Grefenstette, 1991; Koza, 1991) have alternatively emphasized higher-level symbolic rule representations and concentrated on the design of knowledge recombination and mutation operators for manipulating these representations. Our approach adopts the more commonly used fixed-format approach to representing decision models.

Along a second dimension, learning of rule models with GAs has been considered within two basic paradigms: the so-called Classifier Systems or "Michigan" approach (e.g., Holland, 1986), and the so-called "Pitt" approach (e.g., Smith, 1983). At an operational level, these two paradigms can be distinguished by the role given to the GA in the learning process. The Michigan approach assumes a population of individual rules (classifiers) that collectively constitute the current decision model. The GA is given responsibility for generating new rules; the integration of rules into a coherent model is handled by other mechanisms. The Pitt approach, in contrast, is more directly based on the standard GA-based optimization framework in which each individual in the population represents a complete candidate model (i.e., a complete set of rules). Here the scope of the GA is extended to include rule integration. Under the Michigan approach, the GA is typically applied in a controlled fashion, reflecting a bias toward incremental, on-line model development and refinement. A small subset of rules (individuals in the population) are periodically selected for reproduction based on measures of prior rule performance in the task domain, and existing low-performance rules are replaced by the newly generated offspring. Pitt approaches reflect a more off-line training perspective, and have generally relied much more extensively on genetic search. Larger percentages of the population are usually selected (in this case, based on measures of overall model performance), recombined, and replaced on each cycle. Michigan approaches have assumed decision models with a fixed number of rules (the size of the population) whereas Pitt approaches have typically allowed individuals with variable numbers of rules and thus have allowed the final model size to be determined within the search.

The approach to learning rule-based decision models advocated in this article represents a unique synthesis of aspects of both paradigms. Specifically, it combines the off-line and rule integration perspectives of the Pitt approach with the "population-is-model" assumption of the Michigan approach. Our framework has itself "evolved" as a result of our prior experiences in attacking learning from examples problems. Thus, we first summarize the progression of systems that led us to our approach and highlight some of the difficulties that were encountered.

3.2. ADAM

ADAM (Adapative Decision Acquisition Model) was an initial approach to using a GA for inductive classification (Greene, 1987; Greene & Smith, 1987). The problem addressed in this work was a binary discrimination task: to produce a model that reflects consumer buying preference, given a set of examples of past consumer choices (to buy or not buy) in the context of a specific set of product features (e.g., size, color, cost, etc.). One important issue in consumer choice modeling (in addition to producing a model that predicts well) is recovery of the underlying functional form of the consumers' decision model (e.g., is the decision dominated by the value of a specific product feature, or are tradeoffs being made among the values of several feature values?). Since such information is of obvious interest to marketing research, consumer choice modeling is a problem well suited for symbolic induction.

The ADAM system represented a fairly standard adaptation of the Pitt approach to rule learning, with the exception that the special structure of the binary classification problem was exploited to simplify the representation of candidate models. Within ADAM, a model (or individual in the population) was represented as a set of disjunctive clauses that collectively specify the circumstances under which a positive classification (e.g., "buy") should be made. Each clause expressed a particular conjunctive pattern over the entire feature set using a variant of a pattern specification language commonly employed in GA learning research. Specifically, clauses were encoded as fixed-length strings over the alphabet {0, 1, #}, (where, as usual, # designates "don't care") to be matched against a correspondingly ordered binary encoding of the product feature values associated with any given example. For example, in the case of a problem involving three boolean feature detectors d_1, d_2, d_3, the condition *1#0* would be interpreted as

If d_1 *AND* (*NOT* d_3) *Then buy*

If the problem alternatively involved a single eight-valued feature detector d_1, then the interpretation of the same condition might be

If (d_1 = *value*4) *OR* (d_1 = *value*6) *Then buy*

Thus, outcomes were not explicitly represented. If a matching clause was found during application of the model to a given example, then the model's response was interpreted as positive; if no clause was found to match, the response was a negative classification. A similar perspective on model representation, which can be seen equivalently as a collection of "implicit-positive" classification rules, has been exploited by (DeJong & Spears, 1991) in more recent work in GA-based concept learning.

Within the training phase search conducted by ADAM, candidate models were exposed to distinct, overlapping subsets of the overall set of training examples on each generation. This overlapping sample technique was designed to encourage models with appropriately general clauses (since overly specific and overly general models failed to respond properly on successive subsets over time) as well as to provide a computationally efficient evaluation step. The overall fitness of a given model was defined as a weighted sum of predictive accuracy (percentage of correct responses) and model simplicity (number of clauses). Given the sensitivity of these weights to the unknown form of the data-generating function, a simple mechanism for dynamically shifting this bias as the search proceeded was employed. Specifically, increased weight was shifted to model simplicity if and when the number of clauses in the best individual in the population fell below the average number of clauses in the individuals of the randomly generated initial population. A single-point crossover operator, parameterized to operate at either the inter-clause or intro-clause level according to a specified probability, was used as the principal means of generating new candidate models for testing: a clause was randomly selected from each selected parent, the two structures were aligned, and the structures were then crossed at a randomly selected point (at the appropriate level) common to both. The probability of crossing within a clause or at clause boundaries was varied according to a predefined schedule that encouraged more intra-clause exchange early on in the search and more inter-clause manipulation at later stages. The search was terminated after a pre-specified number of generations, and the best model found was then evaluated with respect to a separate holdout set of examples.

In a comparison of symbolic and statistical approaches applied to consumer choice data sets, the ADAM system was found to have superior performance versus a symbolic induction system called CLS (Currim, 1986) (a variant of the ID3 decision tree-building approach) and equivalent performance versus a statistical logit model (the current standard in practice). This comparison was carried out in a factorial design encompassing different noise levels, data-set sizes and generative models. Results demonstrated that ADAM was able to construct decision models that not only predicted well but also captured the functional form of the choice model.

One of the keys to ADAM's success was the strong coupling between its specialized, two-class model representation, which permitted a focused search for the conditions relevant to positive classification only, and the fitness function, which effectively reflected this "optimization" goal and required no finer-level search bias. The cost of these assumptions, on the other hand, was their inability to extend to multi-class problems.

3.3. GARGLE

To explore the application of GAs in more complex, multi-class and continuous valued modeling domains, an experimental testbed called GARGLE (GA-Based Rule Generation Learning Environment) was created (Greene, 1992). While retaining the Pitt approach of manipulating a population of a complete decision models, a new model representation was adopted within GARGLE that replaced ADAM's multi-clause model with a set of ⟨single-clause outcome⟩ rules. This model representation is similar to representations typically exploited in classifier system research (e.g., Wilson, 1987). However, in contrast to the approaches taken in this work, rule outcomes were not explicitly manipulated by the GA in GARGLE. Exploiting the presence of the data set of training examples, a newly created rule was instead assigned an outcome based on the predominant outcome of the set of examples that its condition matches. This approach, referred to as *ex-post assignment*, is illustrated in figure 2.

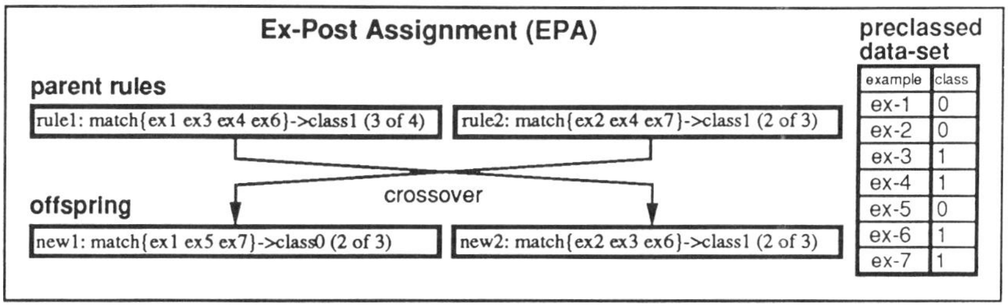

Figure 2. Assigning an outcome to a newly created rule.

Ex-post assignment offers advantages in the context of learning decision models from examples:

- it reduces and simplifies the size of the space to be searched;
- it creates a useful search bias, maximizing the potential utility of any condition that is generated during the search; and
- it provides a common basis for addressing continuous valued as well as discrete valued multi-class decision problems. In the former case, for example, the mean value of the outcomes in the matched example set can serve as the rule's outcome, with the variance providing a measure of rule strength.

While providing the flexibility to address this broader class of decision models, movement to a multi-outcome representation introduced several thorny issues that could be effectively avoided in the simpler two-class approach (many of which are not unknown to research in GA-based learning (Wilson, 1989)) All of these issues related to the problem of creating a search bias that adequately promoted coverage of the training examples while constraining the size of rule sets that were generated. In specifying an appropriate conflict resolution strategy, there was no clearly compatible resolution metric (e.g., accuracy, total matches, specificity, etc.) or combination of metrics from the standpoint of global model-building objectives, nor was it obvious how best to apply resolution criteria. A variety of dominance and consensus strategies (choosing the outcome based on the "best" matching rule or the collective influence of all matching rules, respectively) were tested with mixed results. For example, consensus strategies, in the absence of other biases, tended to promote model growth. Establishment of an appropriate bias toward smaller models within the global fitness measure also proved problematic. While effective in ADAM, the simple weighted use of predictive accuracy and model size as a measure of model fitness invariably led to either unacceptable model growth (too little size bias) or an inability to maintain sufficient diversity to get good coverage of the training set (too much size bias). To compensate, more sophisticated, multi-point crossover operators were explored, which utilized rule coverage similarity measures to bias crossover point selection. While some system configurations did produce fairly reasonable search performance, and several promising mechanisms were developed, the continuing difficulty of simultaneously managing model complexity (size) and maintaining sufficient diversity for good performance within the Pitt approach finally led us to seek an alternative search framework.

4. Coverage-based genetic induction

By starting from the problem and exploiting its inherent structure, a different system organization emerges. This perspective led to the development of COGIN (COverage-based Genetic INduction), a system for symbolic model induction from a set of pre-specified examples. The fundamental organizing principal of the COGIN search framework is the concept of a *coverage-based constraint*. The basic idea applied to modeling a set of examples is to think of the examples as corresponding to niches in an ecology, where the exact number of niches is unknown but presumed to be less than the example set size (i.e., one or more examples per niche). Treating the population as a model and each rule in the model as an individual representing a possible niche, we impose a specific bias in preferring the ecological model, which classifies accurately and requires the fewest niches. This implies a system that will simultaneously search for the ideal number of niches as well as the best individual within each niche.

Consider first the problem of identifying the best individual. To get the highest accuracy with the smallest model, the system needs individuals with high discriminability, that is, individuals that differentiate many examples correctly. By ordering on discriminability, we can identify the "best" individuals. To identify the fewest number of niches, we can move sequentially through this ordering and "allocate" correctly matched examples to each rule along the way. If there are no examples available to a rule when its turn to receive examples comes, it has no unique niche and does not survive. When all the examples have been "consumed," the ecology supports no more rules. The resultant model provides a minimal number of niches with the best available individual designating each niche. Evolution suggests that survival of the fittest will occur where a limited environment causes competitive pressure—it is this coverage-based constraint that provides just such an environment.

4.1. Overview of COGIN approach

Figure 3 depicts the basic search cycle carried out by COGIN to construct a decision model from a pre-specified set of training examples. The system's current model at any point

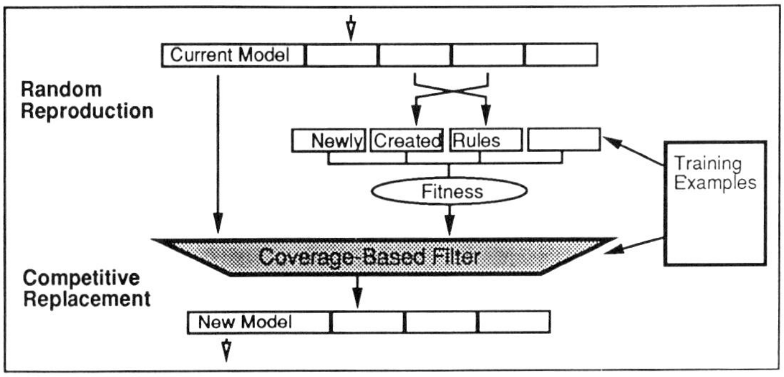

Figure 3. The COGIN search cycle.

during the search is represented as a population of fixed length rules. However, in contrast to typical classifier system approaches, the population *size* (i.e., the number of rules in the model) will vary from cycle to cycle as a function of the coverage constraint that is applied. At the start of each cycle, a percentage of the rules constituting the current model are randomly paired and recombined to produce a set of new candidate rules. Each of these new rules is assigned a fitness measure, based on discriminability with respect to the training set, and then pooled with the original rules of the current model for competitive configuration of a new model. The heart of the cycle is the competition step, which determines the rules that survive to the next generation. This is carried out through application of a *coverage-based filter*, which "allocates" training examples to rules that correctly match them in fitness order. The output of the filter, which constitutes the new decision model, is the set of rules receiving at least one example, and all other candidate rules are discarded. After the new model has been determined, its predictive accuracy relative to the training set is evaluated. If the new model yields a higher predictive accuracy than the best model evaluated thus far in the search, or performs at the same level and contains fewer rules than the previous best model, then a copy is retained off-line as the new best for future comparison and the cycle repeats. The overall search process terminates after a pre-specified number of iterations, and the best model produced during the search is taken as the final model.

4.1.1. An example

Figure 4 provides a representative example of the dynamics of this search process in the context of a two-class problem, assuming a training set of seven examples, with examples {1, 2, and 5} of class 0 and examples {3, 4, 6, and 7} of class 1. Notationally, each rule in the figure is depicted by number, followed by the set of matching examples, its action, and its fitness. For instance, [rl: {1 3 4 6 } → 1(.089)] indicates that rule rl matches examples 1, 3, 4, and 6, assigns class 1, and has a fitness of .089. For our purposes here, entropy is used as a fitness measure; the formulation is provided in Quinlan (1986). (The specific fitness measure employed in COGIN is described later in this section.) At generation N the current model consists of four rules {rl, r2, r3, r4} that correctly predict 6 of the 7 examples (86%) in the training set and have an average rule fitness of .070. Three new rules are created in this cycle (r5, r6, r7), and the "allocation" of training examples within the filter (depicted) leaves no examples for rules r2, r3, or r6. These rules are discarded, and the remaining rules (r7, r5, rl, r4) become the generation $N + 1$ model. Notice that rule r4 survives despite its low fitness, by being the only rule that matched example 7 *correctly*. Notice, too, that this model has the same number of rules as the generation N model but actually a lower predictive accuracy. At the same time, the average rule fitness has increased, suggesting the emergence of better model components. After generating new rules and applying the filter, at generation $N + 2$ the new model is in fact seen to correctly classify all training set examples. Although this example was created for illustrative purposes, this behavior is quite representative of actual system performance. We will return to the issue of search dynamics after summarizing the principal elements of the COGIN inductive framework.

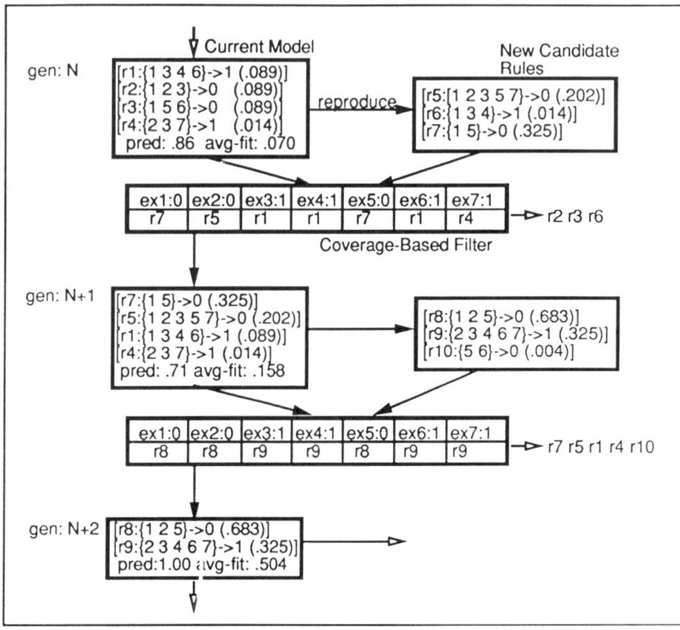

Figure 4. An example of COGIN's search procedure.

4.1.2. Model representation and interpretation

As previously stated, decision models are represented within COGIN as a set of fixed-length rules (the number of which is determined during the search process). The COGIN rule representation is a simple extension of the rule representation employed in GARGLE. Each rule similarly specifies a single conjunctive pattern in the ternary alphabet {0, 1, #} over an ordered binary encoding of the feature set. The extension is the additional association of a negation bit with each feature subpattern, which, if set, designates the complement of the specified value set (the negation bit is ignored in the case where the feature pattern matches all possible values, i.e., "#. . #"). The system is parameterized to run with or without this extension. Thus, for example, if we assumed use of the negation extension in the case of a four-class problem with ordered feature detectors d_1 and d_2 taking on eight and four possible values, respectively, the COGIN rule

$$0\ 11\#\ 1\ 01 \rightarrow 11$$

would be interpreted as

$$(d_1 = value6\ OR\ value7)\ AND\ (NOT\ (d_2 = value1))\ Then\ class3$$

During the application of a model to classify a given example, rule fitness (discussed below) is used as a basis for conflict resolution. In situations where no matching rules are found (i.e., in classifying examples not seen during training), a default classification is made based on the majority class of the training set examples.

4.1.3. Initializing the search

An initial model is constructed by randomly generating rules (using 50% don't-care bits) until all examples in the training set have been correctly matched. Thus, the search is initialized with a random model of the domain in which all niches are covered. The competitive replacement step of the search cycle will enforce this constraint throughout the search, ensuring that no niche description will be lost until replaced by better niche description (i.e., a superior rule).

4.1.4. Generating new candidate rules

New candidate rules are generated on any given cycle of the search by randomly recombining rules in the current model (population). The current rule set is duplicated, and a fixed percentage of parent rules are randomly selected from each copy and paired for reproduction—since all offspring and parents are retained, it is not clear that there are any advantages in using less than 100%. Rules are chosen for participation without replacement, and pairings are made to ensure distinct parents. For each set of paired rules, a single crossover point is randomly selected among the bits in the condition strings (i.e., the crossover point may fall within or between feature patterns), and the condition segments to the right of the crossover point are swapped to produce two new rule condition patterns.[3] An outcome is assigned to each newly created rule using the *ex-post assignment* scheme discussed in section 3. That is, the set of training examples matched by each new rule condition is collected, and the majority outcome of this match set is assigned (choosing randomly in the case of a tie). This match set is recorded with the new rule for subsequent use in assigning fitness.

4.1.5. Assigning rule fitness

As indicated above, the overall goal of producing the smallest possible model with the best possible predictive accuracy requires rules with high discriminability. One well-supported measure of discriminability is the information gain or mutual information (Quinlan, 1986). The information gain tells us how many bits of information about the correct classifications of the examples are provided by a given rule. Consider a set S of examples. Let p_i be the proportion of examples in S belonging to class i. Then the *entropy*, $H(S)$, is generally defined as $-\Sigma_{i=1}^{n} (p_i \, ln(p_i))$. Now consider a rule R. Let *Matched* be the set of examples that R matches, and *Unmatched* be the set of examples that R does not match. With these definitions, the information gain of rule R is

$$Info(R) = H(S) - \frac{[|Matched|\ H(Matched) + |Unmatched|\ H(Unmatched)]}{|S|}$$

Entropy-based measures have been frequently utilized as a search bias in symbolic learning contexts (e.g., serving as the basis for tree splitting in ID3). *Info*(R) is similarly adopted as a basic measure of rule fitness in COGIN.

However, unlike our earlier simplified example (figure 4), the fitness measure assigned to a rule within COGIN is not based solely on its entropy value. In practice, the use of entropy alone creates a bias toward quantity in the sense that it will favor a rule with many matches as long as the number of incorrect classifications is proportionately small, and this characteristic can adversely affect the model-building goal of predictive accuracy. To mediate this bias toward overly general rules, we formulate rule fitness in COGIN as *lexicographically-screened entropy*, defined as

$$Fitness(R) = Lex(R) + Info(R)$$

where *Info*(R) was described previously and takes on values between 0 and 1, and *Lex*(R) is an integer-valued screening function that measures classification accuracy. A user-specified "interval" parameter i is used to establish misclassification error levels (or bins) that are considered equally fit, and *Lex*(R) is defined as

$$Lex(R) = \left\lceil \frac{N}{i} \right\rceil - \left\lceil \frac{Misclass(R)}{i} \right\rceil$$

where N is the total number of examples in the training set, and *Misclass*(R) is the number of misclassifications made by R. For example, if $i = 2$, then the fitness scores assigned to rules r1 through r7 in our previous example of figure 4 (where $N = 7$) are as shown in table 1. The principal effect of the *Lex* term in this case is to lower the competitive priority of r5, making it less likely to survive into the next generation model.

Table 1. Fitness assignment under lexicographically screened entropy.

Rule	Info(R)	Misclass(R)	Fitness(R)
r7: $\{1\ 5\} \rightarrow 0$	.325	0	3.325
r1: $\{1\ 3\ 4\ 6\} \rightarrow 1$	.089	1	3.089
r2: $\{1\ 2\ 3\} \rightarrow 0$	.089	1	3.089
r3: $\{1\ 5\ 6\} \rightarrow 0$	.089	1	3.089
r4: $\{2\ 3\ 7\} \rightarrow 1$	.014	1	3.014
r6: $\{1\ 3\ 4\} \rightarrow 1$	.014	1	3.014
r5: $\{1\ 2\ 3\ 5\ 7\} \rightarrow 0$	.202	2	2.202

There are several more general points to make about the approach to fitness assignment represented by the above calculation.

- The goal of fitness assignment in COGIN is to establish a rank ordering of candidates; for example, the fact that *fitness*($R6$) > *fitness*($R5$) in table 1 is important, not the distance between *fitness*($R6$) and *fitness*($R5$).
- The approach taken here deviates from the approach previously taken in ADAM and GARGLE, where weighted compensatory fitness functions were utilized. The above described "lexicographic" approach does not presume a smooth tradeoff between contributing fitness measures, but instead is structured so that the influence of *Lex*(R) will dominate (or impose constraints on) the influence of *Info*(R). This choice was made to allow the effects of search bias to be more easily controlled and understood. Within *Lex*(R), increasing values of i define increasingly broader (looser) constraints on the influence of *Info*(R).
- A number of alternative measures have been proposed as a means of counterbalancing the undesirable aspects of a search bias based solely on entropy (e.g., statistical tests of significance). We have chosen misclassification error for purposes of simplicity in our current implementation, and make no claims that this is the best choice.

4.1.6. Competititve replacement

While reproduction is random, replacement is not. Newly created rules compete with the original rules in the current model for entry into the population. This is accomplished by a filtering process that utilizes the set of training examples to impose a coverage-based constraint on the size of the new model that is being configured for the next generation of the search. The filtering process proceeds as follows. The entire pool of candidate rules is ordered by fitness. Next, the set of training examples is passed over this list, removing those examples correctly matched by the current rule from further consideration at each step. The subset of rules remaining when the set of training examples becomes empty are eliminated, and the new model is established. Given the use of entropy as a determinant of fitness, the model's rules naturally form a hierarchy with generalist rules dominating and more specific "exception" rules filling in remaining gaps. This suggests an interesting parallel to the "default hierarchies" sought in classifier system research.

4.2. Search bias

At first glance, it would appear that the commitment to random selection of rules for reproduction would disrupt the ability of the search to efficiently allocate trials to high-performance building blocks as suggested by the classical GA's theoretical foundation. However, this does not seem to be the case. In terms of the classical select-reproduce-evaluate GA search framework, we can see application of the coverage-based filter as a form of rank selection (Baker, 1985), in this case with a dynamically determined cutoff on the individuals allowed to participate in influencing the next generation. Selection, in the classical sense, is going on with respect to the larger pool of current and newly created rules on any given cycle. In Eshelman (1991), a very similar search paradigm is defined and analyzed in the context of function optimization and is shown to retain the fundamental property of implicit parallelism.

Moreover, it is not clear what further useful search bias could be introduced to guide reproduction, since rules exist in the model because they fill different niches. One possibility that could be explored is a pairing strategy based on overlap of training examples matched. However, given that replacement will provide competitive pressure, reproduction should therefore emphasize diversity, and this is already promoted through random pairing. One useful bias with respect to uncovering or better filling specific niches is the use of ex-post assignment. As mentioned earlier, this rule-building heuristic maximizes the potential of any rule condition that is generated.

Figure 5 gives some insight the dynamics of COGIN's search. It tracks four values over time, averaged over five runs of one of the conjunctive four-class problem considered in the experimental study described in section 5: the accuracy on the training set, the accuracy on the holdout set (of course unavailable to the system during the search), the average fitness of the rules in the current model, and the number of rules in the current model (the curves of the last two values are scaled to the maximum value seen). We can see that 100% accuracy on the training set is achieved very early on in the search, with the size of the current model generally increasing as well (a function of the dominant accuracy bias). At this point, the entropy bias takes over and focuses the search toward more general rules and correspondingly smaller models (rule sets).

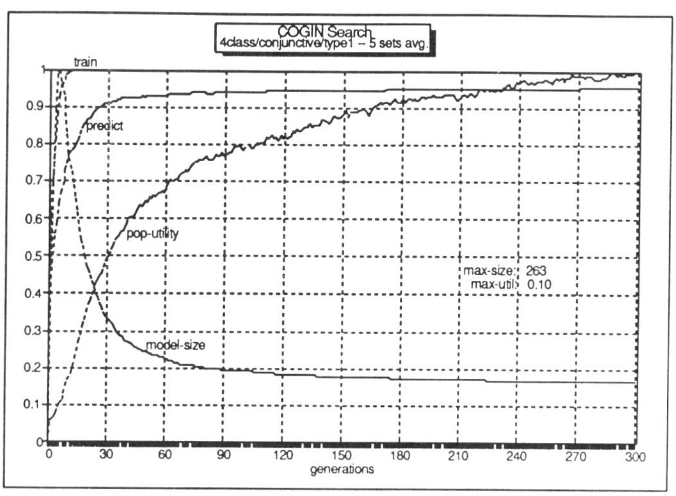

Figure 5. Performance curves for simple conjunctive four-class problems.

4.3. Relationship to other approaches

The use of coverage as an explicit constraint on model (population) size represents a fairly significant departure from previous approaches to the problem of maintaining the diversity in model components (rules) necessary for robust decision-making. Research with Michigan-style classifier systems (the closest match to the COGIN framework) has typically proceeded from the base assumption of a fixed model size, large enough to accommodate the range of decision-making circumstances that the problem solver must face. The generation of higher-performance rules is treated as a separable issue from niche building and is typically handled by separate mechanisms. Use of various measures of individual rule strength to focus selection of rules for recombination (the typical approach) promotes improvement of model performance in niches currently filled by the model. However, this selection bias does little to address the problem of uncovering new niches. Several additional mechanisms have been utilized to provide this latter search bias. The use of partial matching techniques during model interpretation (Booker, 1982) provides the opportunity to accumulate rule strength with respect to a wider range of situations than the rule explicitly implies and thus to focus search toward appropriate expansion of model coverage. A more explicit approach to building new niches has been the use of so-called "triggered operators" (Wilson, 1987; Robertson & Riolo, 1988; Booker, 1989), specialized procedures that are invoked when a non-matched example (or situation) is encountered and that act to construct a new rule through some amount of generalization from this example. Despite the addition of these mechanisms, however, selection according to a rule performance still provides only indirect pressure toward niche building, since rules paired for recombination may be playing distinctly different roles in the model. Recognizing this problem Booker (1989) has advocated the use of restricted mating policies, where only rules that are simultaneously active (i.e., simultaneously match a given situation) during model interpretation are allowed to be paired for recombination, and experimental results have demonstrated the desired effect in the context of concept learning.

What these approaches fail to emphasize, in contrast to the approach taken by COGIN, is the relationship between model complexity and the underlying behavior or decision process that is being modeled. There is an explicit global constraint (the population size), but no finer pressure to cover niches in the most economical fashion. Admittedly, this may be reasonable in the context of developing intelligent agents (where available "brain cells" are the principal constraint). However, as we have previously observed, model simplicity is an important concern in the context of constructing comprehensible decision models from data. Research with Pitt-style rule learning approaches, which has typically manipulated variable-length models, has had to face this issue out of computational necessity. The most common approach has been to simply introduce a bias for smaller models in the fitness function used to estimate the utility of candidate solutions, typically with an underlying global constraint on model complexity (i.e., a maximum allowable number of rules (Smith, 1980). Our prior research with ADAM (Greene, 1987) has demonstrated the sufficiency of this approach relative to binary classification problems. However, as previously discussed in section 3, the problem of diversity and the insufficiency of a size bias to properly address the model performance/complexity tradeoff really only emerge in multi-class contexts. One approach taken to promote diversity within the Pitt approach has been the use of mutlivalued fitness functions (Schaffer, 1984), designed to estimate model performance with respect to each individual class. Other approaches (e.g., Grefenstette, 1990; our explorations with GARGLE) have resorted to similar types of mechanisms as those investigated in classifier system research (e.g., triggered operators, incorporation of individual rule discriminability information into recombination operators). However, neither of these types of extensions directly impact the object of simpler models.

In contrast to this work, COGIN's use of coverage as a global constraint on model complexity, applied in conjunction with a discriminability-based selection/survival bias, provides a conceptual framework that establishes exactly the desired search dynamics for effective induction from examples. Model complexity expands or contracts as necessary to accommodate the data, while fitness based on discriminability of individual rules biases search toward a model of appropriate complexity. Taking the analogy from nature, individuals in the ecology are allowed to intermingle, creating wide variations; but their environment is constrained, and survival depends on being more fit than competitors for the same resource niche.

5. Experimental study

To understand the performance of COGIN in constructing decision models from examples, a two-phase experimental study was designed and carried out. During the first phase, experiments were run to determine appropriate settings for two parameters in COGIN that impact the necessary tradeoff between model predictive accuracy and model size. During the second phase, the results obtained were used as a basis for conducting a comparative performance analysis of COGIN and two symbolic induction systems—CN2 and NewID (a current version of the ID3 methodology developed at the Turing Institute)—across a range of model induction problems. Each problem was represented as a data set of 800 examples, with each example having eight to ten four-valued features and either a two-class or four-class dependent variable (outcome). The data sets were randomly split into 400 training examples (the training set) and 400 prediction examples (the holdout set). To provide a basis for contrasting performance across problems of different complexity, data sets were varied along two dimensions of complexity: noise in the training examples, and the data-generating function.

Along the noise dimension, two levels were considered for each data set: "clean" (i.e., 0%) and "noisy," with 20% random distortion among the features in the training data. Because we were interested in how well each system would recover the true underlying model in the presence of noise (as opposed to evaluating performance on recovery of the "related" underlying model resulting from introduction of noise), the holdout samples were kept clean. With respect to the data-generating function, we considered two types: conjunctive (generally lower complexity) and compensatory (higher complexity). Conjunctive problems are defined by a lexicographic decision rule (e.g., consider feature-X first, then feature-Y, and so on); thus the classification may be determined by a single feature or specific sequence. Compensatory problems are weighted summations of the relevant features; thus a poor value on one feature can be "compensated" by a better value on another. Linear compensatory problems are the raison d'etre for statistical methods but can be problematic for symbolic induction because of the number of tradeoffs that are possible. Conjunctive problems, on the other hand, are difficult for regression approaches but quite consistent with stepwise search. Through experimentation, several conjunctive and compensatory functions were simulated to create instances of varying complexity within both types: from simple two-class to multiple-interaction four-class conjunctive and simpler two-class to multiple feature four-class compensatory. For each data set a copy was then made with and without noise.

As an additional problem, a data set supplied by a sponsoring agency, the U.S. Army Human Engineering Laboratory (HEL), was also used. This HEL data set was of interest because the underlying generating function was based on an important resource resupply problem that had proven difficult in practice for other forecasting methods. It exhibited both conjunctive and compensatory problem characteristics. While reflecting a primarily conjunctive decision process, the training examples contained real-valued situational features, resulting in a pseudo-compensatory behavior. The HEL data set consisted of approximately 1600 examples with seven features of multiple values and a continuous valued outcome, which, for purposes of the experiments described below, was discretized into four classes. Two additional data sets were then generated from this base data set: one with 15% noise in the features only and the other with 15% noise among the discretized outcomes.

5.1. Phase 1: Analysis of COGIN performance characteristics

In the first phase, we evaluated two specific parameters in COGIN expected to be particularly relevant to the goal of generating compact, high-performance models:

- the use of a negation within the individual feature patterns of a rule, and
- the size of the error intervals used in the fitness assignment

The association of a negation bit with each feature pattern provides a richer framework for expressing feature interactions, and thus suggests the possibility of simpler rule models. The error interval in the fitness function acts to tolerate more or less misclassification when selecting the covering set of rules to constitute the population for the next generation. Since it was expected that overfitting might occur under some problem conditions, error interval levels of 1, 2, and 3 were evaluated in combination with the presence or absence of negation in the representation. The results of multiple experiments over several of the above problems indicated a fair amount of insensitivity to these parameter settings. As expected, the use of negation resulted in somewhat shorter rules. Larger error intervals yielded slightly better predictive performance on noisy data, while the smaller intervals performed better in most other cases. It appears that the width of the first interval has the strongest potential influence on the performance. For purposes of assessing the comparative performance of COGIN, CN2, and NewID in phase 2, we selected the parameter values that were most consistent across all problem conditions: the use of a negation bit and a error interval level of 1.

5.2. Phase 2: Comparative performance analysis

The goal of the second set of experiments was to determine the effectiveness of COGIN on problems of varying complexity relative to inductive approaches rooted in a stepwise search framework. CN2 provides two search variants, one designed to construct an ordered rule set and the other operating under the assumption of an unordered set of rules. Since, in either case, CN2 searches a similar space to that searched by COGIN (albeit differently), it provides an interesting opportunity to evaluate how performance varies among these three search orientations, and both versions were included in the evaluation. NewID constructs ordered, decision tree models in a stepwise fashion, and was included in the evaluation to provide an additional point of comparison. Both CN2 and NewID were used with their respective default settings. Because of the possible influence of beam size on CN2, several settings were evaluated. Increasing the beam size had a negligible and, in a number of cases, detrimental effect on performance (possibly due to overfitting). It was decided that the default size of 5 was probably the default for a reason, and hence it was retained. No post-processing of generated models was done for any of the three systems.

As a first-order indicator of effectiveness, the absolute performance levels attained by COGIN, CN2 and NewID relative to each of the problem categories described above can be measured, where performance is defined as predictive accuracy (i.e., the percentage of holdout examples correctly classified.) Our basic conjecture was that GA-based search could effectively address conditions of noise and problem complexity that inhibit stepwise search performance. Since CN2 and NewID have already proven effective under simple, noise-free conditions, our expectation was that as problems became more complex COGIN should begin to outperform them. The range of problem sets was intended to find the conditions where this transition would occur.

Hence, the first hypothesis to be tested was the following:

Hypothesis 1: COGIN should produce decision models with superior predictive accuracy to those produced by CN2 and NewID on problems with greater complexity.

This hypothesis implies that COGIN should do better on at least some of the problems associated with greater complexity. We might also expect that COGIN should perform better (on a relative basis) as the problem became more complex. Therefore, additional insight into the effectiveness of COGIN can be gained by measuring the trend in the performance differential between systems over increasingly more complex problem categories. We would generally expect absolute performance levels to decline with noise and increaisng function complexity (regardless of approach). However, given the apparent greater susceptibility of incremental approaches to interacting features in more complex problems, it was anticipated that, with increasing problem complexity, their relative performance would decrease in comparison to COGIN's.

Hypothesis 2: The relative performance differential between COGIN and CN2 and NewID would become increasingly larger as complexity increases.

As noted in section 2, ordered decision models will tend to be smaller than unordered models because of the inherent if-then-else relationship (which may reduce comprehensibility). While unordered models may rely on an ordered interpretation for resolving conflicts, since their construction does not discourage overlapping coverage of examples, they tend to be larger. Therefore, from the standpoint of the size of models generated by CN2, the most direct comparison to be made is between COGIN and unordered CN2. Although ID3 forms decision trees, they can be roughly interpreted as unordered rule models based on the number of nodes.

In conducting the analysis, 13 problems were defined (each pre-labelled for later reference):

1. [cp0.2cl] compensatory, 2-class (involving 8 features), no noise
2. [cp1.4cl] compensatory, 4-class (involving 6 of 8 features with 2 irrelevant features), no noise
3. [cp1n.4cl] compensatory, 4-class (involving 6 of 8 features), 20% noise
4. [cp2.4cl] compensatory, 4-class (involving 8 features), no noise
5. [cp2n.4cl] compensatory, 4-class (involving 8 features), 20% noise
6. [cj0.2cl] conjunctive, 2-class (involving 8 features), no noise
7. [cj1.4cl] conjunctive, 4-class (involving 8 features), no noise
8. [cj1n.4cl] conjunctive, 4-class (involving 8 features), 20% noise
9. [cj2.4cl] conjunctive, 4-class (involving 10 features), no noise
10. [cj2n.4cl] conjunctive, 4-class (involving 10 features), 20% noise
11. [hel0] hel, 4-class, 8 features, no noise
12. [hel0n1] hel, 4-class, 8 features, 15% noise in features
13. [hel0n2] hel, 4-class, 8 features, 15% noise in classification

Since problems 1 to 10 were represented by simulated data, five instances of each problem were created yielding 50 simulated data sets plus the three hel sets for a total of 53 experiments. Because both CN2 and NewID are deterministic procedures, each was run once per data set using their standard default settings (for CN2, once each for its ordered and unordered search variants). COGIN, using negation and the one-error interval setting, was run four times on each data set using different starting seeds The other principal parameter, the crossover rate, was set at 0.6. Each run was terminated after 300 generations, and the best rule found (based on training performance) was selected and its predictive performance calculated. COGIN's performance on each data set was the average of the multiple runs.[4] Performance on all 53 data sets for all four systems (COGIN, ordered CN2, unordered CN2, NewID) were collected, and composite performances were aggregated for each system based on the 13 problems, the three generation-function types (conjunctive, compensatory, and hel), the two noise levels (0%, 20%), and the six combinations of noise and function type. These results are presented below in tables 2, 3, 4 and 5 (in each table, starred entries indicate the best value for the corresponding problem category). Table 6 indicates the average size of the decision models generated by each system across noise and function type problem categories. In the case of COGIN and CN2, the values given indicate number of rules. In NewID's case, the values indicate the number of decision tree nodes. Because of different hardware platforms and the different developmental stages of the systems tested, no direct time comparisons were made. However, in a companion study (Greene & Smith, 1992), COGIN was found to scale up in roughly the same ratio as ID4 (an automatic pruning version of ID3).

Table 2. Predictive performance of COGIN, CN2, and NewID by problem set: (a) conjunctive problem sets, (b) compensatory problem sets, (c) HEL problem sets.

(a) Predictive Accuracy—Conjunctive Data Sets

	Clean cj0.2cl	Clean cj1.4cl	Noise cj1n.4cl	Clean cj2.4cl	Noise cj2n.4cl
COGIN	0.98*	0.96*	0.91*	0.87*	0.80
Ordered CN2	0.93	0.93	0.89	0.87*	0.82*
Unordered CN2	0.95	0.88	0.85	0.74	0.71
NewID	0.82	0.88	0.84	0.66	0.64

(b) Predictive Accuracy—Compensatory Data Sets

	Clean cp0.2cl	Clean cp1.4cl	Noise cp1n.4cl	Clean cp2.4cl	Noise cp2n.4cl
COGIN	0.83*	0.64	0.59*	0.52*	0.53*
Ordered CN2	0.78	0.56	0.53	0.45	0.44
Unordered CN2	0.78	0.56	0.53	0.45	0.44
NewID	0.64	0.68*	0.58	0.38	0.38

(c) Predictive Accuracy—HEL Data Sets

	Clean hel0	Noise hel0n1	Noise hel0n2
COGIN	0.83*	0.76*	0.66*
Ordered CN2	0.72	0.71	0.60
Unordered CN2	0.74	0.70	0.59
NewID	0.74	0.67	0.55

Table 3. Average performance of COGIN, CN2, and NewID by data generating function type.

	Conjunctive	HEL	Compensatory
COGIN	0.90*	0.75*	0.62*
Ordered CN2	0.89	0.68	0.55
Unordered CN2	0.83	0.68	0.55
NewID	0.77	0.65	0.53

Table 4. Average performance of COGIN, CN2, and NewID on clean vs. noisy training data.

	Clean Examples	Noisy Examples
COGIN	0.80*	0.71*
Ordered CN2	0.75	0.67
Unordered CN2	0.73	0.63
NewID	0.68	0.61

Table 5. Average performance of COGIN, CN2, and NewID by noise and generating function type.

	Clean Conj	Clean HEL	Clean Comp	Noisy Conj	Noisy HEL	Noisy Comp
COGIN	0.93*	0.83*	0.66*	0.85	0.71*	0.56*
Ordered CN2	0.91	0.72	0.60	0.87*	0.66	0.48
Unordered CN2	0.86	0.74	0.60	0.78	0.65	0.49
NewID	0.78	0.74	0.57	0.74	0.61	0.48

Table 6. Average size of COGIN, CN2, and NewID models.

	Clean Conj	Clean HEL	Clean Comp	Noisy Conj	Noisy HEL	Noisy Comp
COGIN	50	209	157	119	275	195
Ordered CN2	33	103	70	38	109	84
Unordered CN2	60	198	117	81	240	134
NewID	159	380	266	221	490	325

5.3. Interpretation of results

Tables 2 through 5 support our initial hypotheses (Hypothesis 1) that COGIN can provide superior performance on more complex problem sets. Less expected was its superior performance on the simpler problems as well. While some of the one-to-one comparisons are not large enough to be significant, the overall consistency is significant and compelling. Further, most of the aggregated comparisons and specifically those between COGIN and unordered CN2 significantly favor COGIN.

As indicated in table 6, the number of rules in the models generated by COGIN and unordered CN2 is similar across all problems, although unordered CN2 models are somewhat more compact in most cases. Also, as indicated and expected, the ordered CN2 models are significantly smaller in most conditions.

Consideration of the second hypothesis stated above, whether genetic search becomes more effective as problem complexity increases, requires an appropriate ordering of the test problems. One difficulty in this respect is the lack of a specifically calibrated complexity measure. A conjunctive data set with considerable interaction may be more complex than a simple compensatory. The addition of noise complicates matters further. Finally, although the hel data share characteristics of both conjunctive and compensatory problems, there are obviously additional dimensions of complexity (e.g., size of the problem) that do not enable us to reliably place it along the conjunctive-compensatory continuum. Overall, the selection of the data sets was intended to provide comparisons on a variety of conditions, not specifically to rank them according to complexity. We therefore use the average performance of all systems tested as a basis for ordering the set of test problems. Computing the average performance of all systems in each function/noise problem category, we get the pattern of problem complexity depicted in figure 6.

clean-conj → noisy-conj → clean-hel → noisy-hel → clean-comp → noisy-comp

Figure 6. Problem categories ordered in increasing complexity.

Table 7. Relative performance differences between COGIN and CN2/NewID by increasing problem complexity.

	Clean Conj	Noisy Conj	Clean HEL	Noisy HEL	Clean Comp	Noisy Comp
COGIN	1.07	1.06	1.10	1.08	1.09	1.11
Ordered CN2	−.03	−.00	−.15	−.08	−.11	−.15
Unordered CN2	−.09	−.09	−.12	−.10	−.11	−.15
NewID	−.17	−.14	−.12	−.15	−.16	−.16

This ordering suggests that the generating function has the strongest influence on complexity, with the presence of noise a secondary contribution. In general, this pattern of categories seems to be a rough approximation of increasing problem difficulty. Table 7 gives an indication of the relative performance differences between systems across this pattern of problem categories. When comparing performance differences, the values shown are normalized within the given category to avoid scaling effects, so entries for COGIN in this table represent normalized predictive performance percentages based on the mean performance of all systems tested. Entries for other systems are the normalized point differences from the COGIN values.

Figure 7 shows a simple regression of CN2 and New-ID values from table 7. Given the caveats on ordering complexity, the data show a solid correlation of 0.57 for ordered CN2 and 0.62 for unordered CN2, with the slope of the regression implying a positive relationship between increasing complexity and the relative performance difference (although we do not presuppose a linear relationship). For NewID the correlation is 0.002, suggesting no relationship. As noted, both the hel data set and the influence of noise confound a reliable ordering of complexity. To evaluate the possibility of a performance trend, we separate the noise and data-function conditions and examine the relative difference in performance

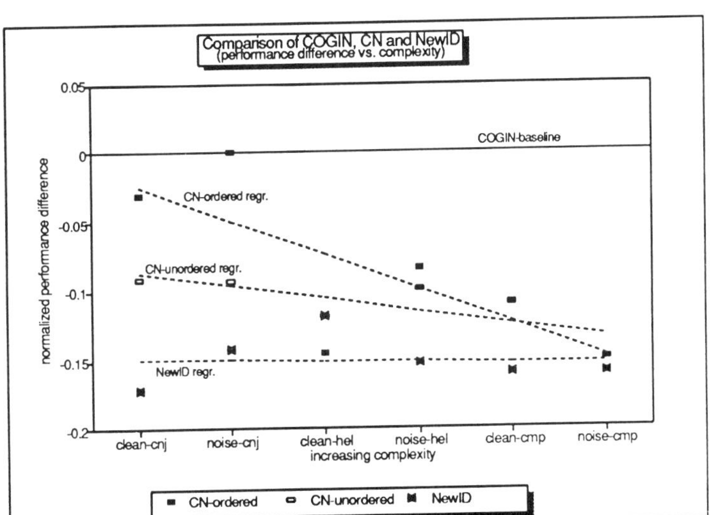

Figure 7. Trend analysis.

Table 8. Relative performance differences between COGIN and CN2 by clean vs. noisy training data.

	Clean	Noisy
COGIN	1.09	1.09
Ordered CN2	−.08	−.05
Unordered CN2	−.10	−.09
NewID	−.16	−.14

between the noise and data-function conditions and examine the relative difference in performance between COGIN and the other systems. The first comparison, table 8, looks at whether COGIN's performance advantage increases as noise increases.

From the table, it is obvious that although the performance is significantly better in both conditions, there is no apparent trend advantage with respect to noise. Examining pairwise comparisons of problem sets confirmed that results were mixed. That is, the performance difference with respect to noise increased for some problems and systems and decreased or remained the same for others—with no obvious pattern. One probable factor is that the dominant influence of the data-generating function overshadows the effect of noise. As noted with the regression, the number of data points does not support breaking the categories further. Given that the data function is a stronger correlate of complexity, to best test the hypothesis of a performance *trend* we examine the relative performance differences between COGIN and the alternative systems based on conjunctive versus compensatory problems as shown in table 9 below.

Given that the dominant complexity effect is the data-generating function, the clearest comparison we can use with confidence is the aggregates of the compensatory and conjunctive experiments, allowing us to formulate hypothesis 2 as a T-test of the difference between two means. We consider the case of ordered CN2 first. If M1 is the mean of the difference between COGIN and ordered CN2 on all conjunctive problem sets, and M2 similarly for all compensatory problem sets, we have the null hypothesis, [H_0: M2 − M1 = 0] and the alternate hypothesis, [H_1: M2 − M1 > 0]. Given 10 problem sets and 8 degrees of freedom, we determine a T-value of 4.309, which rejects H_0 and supports the positive difference at .9999% confidence—in strong support of our second hypothesis. With the data available and using the same tests on unordered CN2 and NewID, the T-statistics are 0.43 and −.25, respectively. We cannot reject the null that there is no clear trend versus

Table 9. Relative performance differences between COGIN and CN2 by data generating function type.

	Conjunctive	HEL	Compensatory
COGIN	1.07	1.09	1.10
Ordered CN2	−.02	−.11	−.10
Unordered CN2	−.09	−.11	−.09
NewID	−.16	−.14	−.14

complexity. Nonetheless, the results suggest an overall normalized advantage of COGIN's competitive-style search over the stepwise feature searches of unordered CN2 and NewID of 10% and 15%, respectively, on this problem set.

5.4. Summary

The purpose of this exploratory study was to evaluate the performance of COGIN over a range of problem conditions emphasizing complexity. Phase 1 examined several parameter variations for COGIN and, based on consistency, settled on the use of a negation bit and a one-error interval level. Phase 2 contrasted how three systems, COGIN, CN2, and New-ID, performed across a variety of problems with different combinations of noise and complexity. Results not only confirmed our initial hypothesis that COGIN would perform better on more complex problems, but also indicated that the GA-based system performed better on simpler problems as well. We then examined whether the relative performance gap between COGIN and the other tested systems actually widened with increasing problem complexity. Significance tests supported this claim in the case of the ordered search variant of CN2, but no significant trends relative to unordered CN2 and New-ID were found.

While these results are promising, three points are worth noting. First, a significant amount of research has investigated the use of various pre- and post-processing procedures to tune and enhance the performance of both CN2 and New-ID. In conducting the above experimental study, both systems were evaluated in their default states, and none of the systems tested (including COGIN) was permitted the luxury of such auxiliary mechanisms. Second, the experimental design executed above was exploratory in nature, and more comprehensive attempts to characterize problem complexity and analyze comparative performance along this dimension are clearly required. One important dimension of complexity that was not emphasized in the above study and clearly needs to be investigated is size in terms of the problem, the sample, and scalability. In a companion study, however, we have shown the basic COGIN system to perform favorably versus extensively post-processed decision-tree models with respect to both prediction and scalability (Green & Smith, 1992). Finally, this study has focused on evaluating the effectiveness of COGIN in relation to traditional induction techniques. Further studies are needed to examine the comparative performance of COGIN and other GA-based learning frameworks.

6. Discussion

In this article, we presented the COGIN framework for induction of decision models from examples. In contrast to previous research in GA-based learning, the COGIN framework exploits the special character of this class of problems, and was designed to directly address the model performance and model complexity tradeoff that this problem presents. Given this emphasis, COGIN represents a fairly significant departure from standard GA-based learning paradigms. Key in this regard is the shift from a "reproduction of the fittest" to a "survival of the fittest" perspective, based more directly on an ecological model where the environment imposes a capacity constraint and survival depends on an ability to exploit a particular niche. Viewing the set of training examples as being representative of specific niches to be described by model rules, coverage is used as an explicit basis for constraining model complexity (size) while at the same time promoting the diversity required to produce good model performance.

COGIN is still very much an evolving system, and several aspects of its design are under active investigation. One issue involves gaining a better understanding of the criteria that should drive both the ordering of rules during competitive replacement and the resolution of conflicts during model interpretation. In the current implementation, a lexicographic fitness function is used to combine entropy and accuracy metrics, relying on an error interval to mediate the potentially harmful effects of strict reliance on entropy. However, the COGIN search framework can flexibly accommodate a variety of strategies, both lexicographic and compensatory, for combining various fitness-related measures (including those commonly applied in other research in GA-based learning). The tradeoffs between various ordering strategies within this framework are still very much an open question. A related area of interest is the introduction of dynamic control of the influence of various parameters during the search. This concept of shifting bias was effectively (albeit simplistically) exploited in the earlier ADAM system, and our current system offers far richer opportunities.

Another principal direction of our current research concerns generalization of the COGIN framework to accommodate modeling problems that require continuous-valued classification. The COGIN approach to dynamic niche allocation based on coverage appears to provide a much stronger framework for exploiting the concept of ex-post assignment than did the search architecture of our earlier GARGLE system. For example, variance in the outcomes of matched examples provides a natural analog to accuracy in the current lexicographic approach to fitness assignment. Recent work reported in Packard (1990) also suggests several interesting ways of measuring and incorporating continuous values, which could work very effectively in COGIN. Such continuous-valued model-building capabilities are an important extension of current symbolic induction systems.

The goal of this research was to develop an induction system that could exploit the potential of a genetic algorithm to overcome the limitations inherent in stepwise search techniques. While preliminary, the experimental results reported in this article clearly indicate the comparative robustness of COGIN's search framework. The development of accurate, comprehensible decision models for real-world problems requires systems that can effectively navigate complex problem spaces. In this regard, we believe COGIN offers a viable alternative to current inductive systems.

Acknowledgments

This work was supported in part by the US ARMY Human Engineering Laboratory under contract DAAD05-90-R-0354, and the Robotics Institute. The authors would like to thank the Turing Institute for providing us with the CN2 and NewID software and Robin Boswell for his assistance in their use. We would also like to thank Rick Riolo, John Grefenstette, and two anonymous reviewers for helpful comments on an earlier draft of this article.

Notes

1. Nonhomogeneity refers to the variables having different relationships in different parts of the measurement space (Breiman, 1984).
2. In genetics, this means a nonreciprocal interaction between nonalternative forms of a gene in which one gene suppresses the expression of another affecting the same part of an organism.
3. Our decision to employ a single-point crossover operator was motivated strictly by reasons of historical bias and computational simplicity. Multi-point and uniform (Syswerda, 1989) crossover operators are equally applicable within the COGIN framework, and we plan to evaluate their comparative utility.
4. The actual variance in performance across runs was very low, usually less than 1%.

References

Baker, J.E. (1985). Adaptive selection methods for genetic algorithms. *Proceedings First International Conference on Genetic Algorithms and their Applications*. Pittsburgh, PA: Morgan Kaufmann, pp. 101–111.

Booker, L. (1989). *Intelligent behavior as an adaptation to the task environment*. Doctoral dissertation, Department of Computer and Communication Sciences, University of Michigan, Ann Arbor, MI.

Booker, L. (1989). Triggered rule discovery in classifier systems. *Proceedings 3rd International Conference on Genetic Algorithms*. Fairfax, VA: Morgan Kaufmann, pp. 265–274.

Breiman, L., Friedman, J., Olshen, R., & Stone, C. (1984). *Classification and regression trees*. Monterey, CA: Wadsworth.

Clark, P., & Niblett, T. (1989). The CN2 induction algorithm. *Machine Learning*, *3*(4), 261–283.

Currim, I., Meyer, R.J., & Le, N. (1986). *A concept learning system for the inference of production models of consumer choice*. Working paper, UCLA.

DeJong, K.A., & Spears, W.M. (1991). Learning concept classification rules using genetic algorithms. *Proceedings 12th International Conference on Artificial Intelligence*. San Mateo, CA: Morgan Kaufmann.

Eshelman, L. (1991). The CHC Adaptive Search algorithm: How to have safe search when engaging in nontraditional genetic recombination. In G.J.E. Rawlins (Ed.), *Foundations of genetic algorithms*. San Mateo, CA: Morgan Kaufmann.

Goldberg, D. (1985). Dynamic systems control using rule learning and genetic algorithms. *Proceedings of the Ninth International Joint Conference on Artificial Intelligence (Los Angeles, California)*. Morgan Kaufmann, pp. 588–592.

Goldberg, D.E. (1989). *Genetic algorithms in search, optimization and machine learning*. Reading, MA: Addison-Wesley.

Greene, D.P. (1987). Automated knowledge acquisition: Overcoming the expert systems bottleneck. *Proceedings of the Seventh International Conference on Information Systems*. Pittsburgh, PA: Lawrence Erlbaum Assoc., pp. 107–117.

Greene, D.P. (1992). *Inductive knowledge acquisition using genetic adaptive search*. Doctoral dissertation, The Graduate School of Industrial Administration, Carnegie Mellon University, Pittsburgh, PA.

Greene, D.P., & Smith, S.F. (1987). A genetic system for learning models of consumer choice. *Proceedings of the Second International Conference on Genetic Algorithms and their Applications*. Boston, MA: Morgan Kaufmann, pp. 217–223.

Greene, D.P., & S.F. Smith. (1992). COGIN: Symbolic induction with genetic algorithms. *Proceedings Tenth National Conference on Artificial Intelligence*. San Mateo, CA: Morgan Kaufmann.

Grefenstette, J.J., Ramsey, C.L., & Schultz, A.C. (1990). Learning sequential decision rules using simulation models and competition. *Machine Learning*, *5*(4), 355–381.

Grefenstette, J.J. (1991). Lamarkian learning in multi-agent environments. *Proceedings 4th International Conference on Genetic Algorithms*. San Mateo, CA: Morgan Kaufmann, pp. 303–310.

Holland, J.H. (1975). *Adaptation in natural and artificial systems*. Ann Arbor, MI: University of Michigan Press.

Holland, J.H. (1986). Escaping brittleness: The possibilities of general purpose learning algorithms applied to parallel rule-based systems. In R. Michalski, J. Carbonell, and T. Mitchell (Eds.), *Machine learning: An artificial intelligence approach* (Vol. 2). San Mateo, CA: Morgan Kaufmann.

Koza, J. (1991). Evolving a computer program to generate random numbers using the genetic programming paradigm. *Proceedings of the Fourth International Conference on Genetic Algorithms and their Applications*. San Diego, CA: Morgan Kaufmann, pp. 37–44.

MacDonald, B.A., & Witten, I.H. (1989). A framework for knowledge acquisition through techniques of concept learning. *IEEE Transactions on Systems Man and Cybernetics*, *19*(3), 499–512.

Michalski, R.S., & Chilauski, R. (1980). Learning by being told and learning from examples. *International Journal of Policy Analysis and Information Systems*, *4*, 210–225.

Packard, N.H. (1990). A genetic learning system for the analysis of complex data. *Complex Systems*, *4*. 543–572.

Quinlan, J.R. (1984). Inductive inference as a tool for the construction of high-performance programs. In R. Michalski, J. Carbonell, and T. Mitchell (Eds.), *Machine learning: An artificial intelligence approach* (vol. 1). Palo Alto, CA: Tioga Press.

Quinlan, J.R. (1986). The effect of noise on concept learning. In R. Michalski, J. Carbonell, and T. Mitchell (Eds.), *Machine learning: An artificial intelligence approach* (Vol. 2). San Mateo, CA: Morgan Kaufmann.

Robertson, G.C., & Riolo, R.L. (1988). A tale of two classifier systems. *Machine Learning*, *3*, 139–159.

Schaffer, J.D. (1984). *Some experiments in machine learning using vector evaluated genetic algorithms*. Doctoral dissertation, Department of Electrical Engineering, Vanderbilt University, Nashville, TN.

Smith, S.F. (1980). *A learning system based on genetic adaptive algorithms*. Doctoral dissertation, Department of Computer Science, University of Pittsburgh, Pittsburgh, PA.

Smith, S.F. (1983). Flexible learning of problem solving heuristics via adaptive search. *Proceedings 8th International Joint Conference on AI*.

Syswerda, G. (June 1989). Uniform crossover in genetic algorithms. *Proceedings of the Third International Conference on Genetic Algorithms and their Applications*. Fairfax, VA: Morgan Kaufman.

Wilson, S.W. (1987). Classifier systems and the animat problem. *Machine Learning*, *2*, 199–228.

Wilson, S.W., & Goldberg, D.E. (1989). A critical review of classifier systems. *Proceedings 3rd International Conference on Genetic Algorithms*. Fairfax, VA: Morgan Kaufmann, pp. 244–255.

Received November 5, 1991
Accepted January 23, 1992
Final Manuscript October 23, 1992

Machine Learning, 13, 259–284 (1993)

Genetic Reinforcement Learning for Neurocontrol Problems

DARRELL WHITLEY, STEPHEN DOMINIC, RAJARSHI DAS, AND
CHARLES W. ANDERSON WHITLEY@CS.COLOSTATE.EDU
Computer Science Department, Colorado State University, Fort Collins, CO 80523

Abstract. Empirical tests indicate that at least one class of genetic algorithms yields good performance for neural network weight optimization in terms of learning rates and scalability. The successful application of these genetic algorithms to supervised learning problems sets the stage for the use of genetic algorithms in reinforcement learning problems. On a simulated inverted-pendulum control problem, "genetic reinforcement learning" produces competitive results with AHC, another well-known reinforcement learning paradigm for neural networks that employs the temporal difference method. These algorithms are compared in terms of learning rates, performance-based generalization, and control behavior over time.

Keywords. Genetic algorithms, reinforcement learning, neural networks, adaptive control

1. Introduction

Recent applications of genetic algorithms to neural network weight optimization produce results roughly competitive with standard back propagation (Montana & Davis, 1989; Whitley, Starkweather, & Bogart, 1990c); but at the same time, there are supervised training paradigms that are significantly faster than back propagation. The application of genetic algorithms to the evolution of neural network topologies has also produced interesting results, but has been largely limited to small problems (Harp, Samad, & Guha, 1990; Whitley & Bogart, 1990a; Schaffer, 1990; Miller & Todd, 1989). There are domains, however, where genetic algorithms can make a unique contribution to neural network learning. In particular, because genetic algorithms do not require or use derivative information, the most appropriate applications are problems where gradient information is unavailable or costly to obtain. Reinforcement learning is one example of such a domain.

The goals of this article are twofold. Our first goal is to demonstrate how genetic algorithms can be used to train neural networks for reinforcement learning and "neurocontrol" applications (Werbos, 1989). As background, we review genetic algorithms and discuss the general problem of optimizing neural network weights using genetic algorithms. A variant of the genetic algorithm is used to optimize the weights of a neural network for the simulated control task of balancing an inverted pendulum. We have chosen this problem because it has been studied using a number of different approaches. The second goal of this article is to compare *genetic reinforcement learning* against another well-known method for training neural networks for reinforcement learning, the "Adaptive Heuristic Critic" (AHC) algorithm (Barto, Sutton & Anderson, 1983; Sutton, 1988; Anderson, 1987). The Adaptive Heuristic Critic (AHC) belongs to a class of adaptive critic reinforcement-learning algorithms that rely upon both a learned evaluation function and a learned action function. Adaptive critic algorithms are designed for reinforcement learning with delayed rewards. The AHC algorithm uses the temporal difference method to train an evaluation network that learns to predict failure. The prediction is then used to heuristically generate plausible target outputs at each time step, thereby allowing the use of back propagation in a separate neural network that maps state variables to output actions. The comparisons offered in this article include not only learning rates but also performance tests designed to measure generalization, as well as graphs displaying the control behavior over time of specific individual networks in specific situations.

Although not often thought of in this way, genetic algorithms are, in a sense, inherently a reinforcement learning technique; the only feedback used by the algorithm is information about the relative performance of different strings representing different potential solutions. Genetic algorithms and reinforcement learning have been linked in the past, but the connection has largely been in the context of classifier systems. Classifier systems use genetic algorithms to optimize and discover simple pattern matching rules using a production system paradigm (Holland, 1986; Goldberg, 1989; Booker, Goldberg, & Holland, 1989). Grefenstette (1989, 1990, 1991) has also used genetic algorithms to optimize and discover sets of (symbolic) rules for sequential decision-making tasks. In the work reported here, genetic algorithms are not used for training classifier systems or other rule-based systems, but rather to train a neural network using only sparse feedback from the environment concerning performance.

Section 2 of this article briefly reviews genetic algorithms and highlights the major issues related to applying genetic algorithms to neural network weight optimization. The inverted pendulum problem is outlined in section 3, while section 4 delineates the genetic reinforcement learning paradigm. A short description of the Adaptive Heuristic Critic algorithm is presented in section 5. Results of comparative tests between the two approaches are reported in section 6.

2. Background: Genetic algorithms and neural networks

Traditionally, genetic algorithms search by manipulating a population of binary strings that represent different potential solutions, each corresponding to a sample point from the search space. The initial population is randomly generated. New sample points are generated by recombining information from two parent strings drawn from the current population. Consider the following binary strings: 1101001100101101 and yxyyxyxxyyyxyxxy, where x and y are used to represent 0 and 1. Using one or more randomly chosen "crossover points," recombination could occur as follows:

```
11010 \ / 01100101101  →  11010yxxyyyxyxxy
yxyyx / \ yxxyyyxyxxy  →  yxyyx01100101101
```

Another key element of genetic algorithms is the use of selective pressure to allocate reproductive trials. By allowing strings representing above-average solutions in the current population to "reproduce" more often than strings representing below-average solutions, the genetic algorithm will allocate more trials to regions in hyperspace that tend to contain above-average solutions. Genetic algorithms are capable of performing a global search of a space because they can rely on hyperplane sampling to guide the search instead of searching along the gradient of a function.

When the string encoding is binary, the search space can be modeled as a binary hypercube. Different regions in hyperspace are represented by schemata. For example, the schema 0**1****** represents an "order-2" hyperplane, since two bits are specified in the schema. The "*" symbol represents a "don't care" operator. This hyperplane contains 25% of all strings in the search space: all strings with a 0 as the first bit and a 1 as the fourth bit. Recombination generates new sample points while preserving hyperplane information through inheritance. For example, if 10101100 and 11011110 are recombined without mutation, both parents and the offspring reside in the hyperplane 1***11*0. The fundamental theorem of genetic algorithms (Holland, 1975; Schaffer, 1987; Goldberg, 1989) demonstrates how it is possible to sample individual hyperplans and allocate either an increased representation in the population if they represent regions of above-average fitness or decreased representations for regions of below-average fitness. That the genetic search efficiently samples numerous hyperplanes in parallel is referred to as *implicit parallelism*. Theoretically, the result is a robust search method capable of searching nonlinear multimodal functions without using gradient information.

In all of our experiments, we use a version of the genetic algorithm we refer to as the GENITOR algorithm. The mechanics of the algorithm are as follows. A population of strings is randomly generated. Each string is evaluated and the population is sorted by ranking strings in terms of their evaluations. A random selection function with a linear bias towards the higher-ranked strings is used to stochastically choose two parents for recombination. The parents are recombined so as to produce a single offspring (i.e., two offspring are generated and one is randomly discarded). After the offspring is evaluated, it replaces the lowest-ranked string in the population and is inserted into its appropriate rank location. The differences between this and other genetic algorithms, as well as their comparative performance on several optimization problems, are discussed in Whitley and Kauth (1988) and Whitley and Hanson (1989).

Several researchers have attempted to apply genetic algorithms to neural network weight optimization problems. The most straightforward way of doing this is to encode each weight in the network as a binary substring; the entire binary encoding would be a string composed of these concatenated substrings. In our previous experiments using genetic algorithms with binary encodings, we have found that genetic algorithms that reply on recombination and limited mutation easily solve small neural network problems such as exclusive-or and 4-2-4 encoders. Given large populations (5000 strings) and a large number of recombinations (e.g., 1 million) we have been able to optimize more complex networks such as a network to add two two-bit numbers using four hidden units and a coding of 280 bits (eight bits per weight). Nevertheless, the resulting search time is dramatically slower than back propagation, and we have not been able to scale these results to significantly larger problems (Whitley & Starkweather, 1990b).

Our experiments suggest that one problem with genetic algorithms for larger neural network problems is that multiple symmetric representations exist for any single neural network. Recombining encodings for functionally dissimilar neural networks can result in inconsistent feedback to the genetic algorithm in the form of inconsistent hyperplane samples. In other words, recombining two good but dissimilar networks can result in offspring that are not viable. For example, consider two small neural networks with two hidden nodes each. Both the networks may successfully learn to compute the same function mapping, but the two solutions may not be compatible because one network learns to compute subfunction A in hidden node $H1$ and subfunction B in hidden node $H2$, while the other network learns just the opposite assignment of functionality to the hidden units. Recombining functionally dissimilar strings will tend to produce offspring that redundantly represent some hidden nodes (functionality) and fail to represent the functionality of other hidden nodes. It is also not just the assignment of functionality to hidden units that may be different. The functionality can also be different in the sense that the sets of weights corresponding to two neural networks can also be scaled differently. Again, the recombination of networks with dissimilar sets of weights can result in offspring with poor performance. We refer to the problem of recombining dissimilar networks as the *structural/functional mapping problem* (Whitley et al., 1990c).

Some basic changes in the GENITOR algorithm resulted in the ability to handle relatively large neural net training problems; the results we have obtained on supervised learning problems (Whitley et al., 1990c) are similar to those reported by Montana and Davis (1989). Only three major implementation differences exist between the algorithms that have failed to optimize larger networks and those we have used to produce positive results. First, the problem encoding is a real-valued instead of binary. This means that each parameter (weight) is represented by a single real value and that recombination can only occur between weights. Second, a much higher level of mutation is used; traditional genetic algorithms are largely driven by recombination, not mutation. Third, a small population is used (e.g., 50 individuals). The use of small populations reduces the exploration of the multiple (representationally dissimilar) solutions for the same net. The stronger reliance on mutation also helps to avoid this problem, since no recombination is involved when mutation occurs.

Another way of trying to deal with the structural/function mapping problem is to try to identify the functionality of each hidden unit and then to swap similar hidden units. However, the experiments of Montana and Davis (1989) suggested there is little difference in performance when using operators that crossover weights or sets of nodes, or operators that attempt to extract hidden unit functionality and swap similar units. Therefore, we merely recombine the weights without attempting to restrict crossover or identify similar hidden units. Thus, our approach does not solve the structural–functional problem directly; instead, it bypasses the problem by using a small population and high mutation rates. Although the implementation details are not so different from conventional genetic algorithms, the empirical evidence suggests that the result is a type of stochastic hill-climbing algorithm that we refer to as a *genetic hill-climber* (Whitley et al., 1991). This algorithm has solved some large supervised learning problems (a neural network for signal detection with approximately 500 weighted connections) in approximately the same amount of time as back propagation (Whitley et al., 1990c).

The details of the real-coded genetic algorithm used in this article are given in figure 1. The use of adaptive crossover and mutation rates is again based on the work of Davis (1989). While it is somewhat unusual to include the crossover and mutation probabilities as an "allele" in the problem encoding, the idea was to have an expedient means of implementing

**** Initialization Phase: For each individual do the following ****

- Set all weights in the network to a random value between ± 2.5. Set one allele representing the probability of crossover to a random value between 0 and 1. Evaluate each individual and sort the population according to the fitness.

**** Iteration Phase ****

(1) Select two individuals according to relative fitness using linear-bias selection. Crossover with probability determined by the crossover probability allele of the string selected as parent 1; otherwise perform mutation on parent 1.

(2) The offspring always inherits the crossover probability of parent 1. If parent 1 has a higher fitness than the offspring, increment the offspring's crossover probability by a factor of 0.10 (to maximum 0.95); otherwise decrease the crossover probability by a factor of 0.10 (to minimum 0.05).

(3) Evaluate the new offspring and insert in the population according to fitness. Continue "Iteration" until error is acceptable or MAX-ITERATIONS = True.

**** Operators ****

- **Mutation:** Mutate all weights on the first selected individual by adding a random value with range ± 10.0.

- **Crossover:** Perform no crossover if the parents differ by two or fewer alleles. Otherwise, recombine the strings using one-point crossover between the first and last positions at which the parents have different weight values.

Figure 1. Implementation details of the real coded genetic algorithm.

adaptive operators. We conjecture that the exact nature of the adaptive operators is not critical as long as it has the effect that the algorithm automatically increases the probability of mutation when the population is converging. Note that crossover and mutation are mutually exclusive operators in the current implementation, so that the probability of crossover decreases as the population converges. We also note that the mutation operator is somewhat extreme: the offspring produced by mutation is a new random point located within a specific radius of the parent string.

While the genetic hill-climber appears to be roughly competitive with back propagation, it fails to be competitive when compared with faster supervised learning methods such as cascade correlation (Fahlman, 1990). Our tests on an adder for two-bit numbers and a large signal pulse detection problem show the genetic hill-climber to be much slower than cascade correlation. Nevertheless, genetic algorithms can make a unique and valuable contribution to the field of neural networks, especially for learning problems when gradient information is unavailable or costly to obtain. Currently, most neural networks are feed forward networks that use sigmoid transfer functions or radial basis functions. These choices make gradient information relatively easy to obtain. However, if one wishes to use other, more complex kinds of transfer units, such as "product units" or even splines—or if one wished to train fully recurrent networks—then computing gradient information becomes far more costly. For these kinds of networks, genetic algorithms may represent a viable alternative. Similarly, training networks in situations where only sparse reinforcement is available is a difficult learning problem because one does not know in advance what the "correct" output of the network should be at each time step; therefore, the derivative information needed to drive gradient descent methods is not directly available. Most current methods for training neural networks using sparse feedback rely on a second network that either heuristically estimates plausible target outputs or that generates targets by back propagating errors through time. Both approaches involve considerable computational overhead.

3. Reinforcement learning for balancing an inverted pendulum

Control systems represent an important application for reinforcement learning algorithms because for some problems it is not feasible to analytically derive controllers; as a result, the training data needed for supervised learning such as simple back propagation may not be directly available. Werbos (1989) provides a brief review of neurcontrol learning paradigms; a more complete discussion is found in Barto (1990) and Werbos (1990). The inverted pendulum problem (which is also sometimes referred to as pole-balancing and cart-centering) is a classic control task. Several researchers have looked at ways of training neural networks for this problem. Of particular relevance are those methods that treat this as a reinforcement learning problem by using algorithms that rely on the relatively uninformative failure signal for feedback, such as Michie and Chambers (1968), Barto, Sutton, and Anderson (1983), Sutton (1988), Selfridge, Sutton, and Barto (1985) and Anderson (1987, 1989).

The inverted pendulum problem involves controlling an inherently unstable mechanical system of a cart and a pole that is constrained to move within a vertical plane. The objective is to keep the pole balanced and to avoid track boundaries. Training a neural network to solve the inverted pendulum problem cannot be directly accomplished using standard supervised training if we wish to use only performance information during training. Supervised training implies that for each input in the training set there is a known desired output. But in this problem we would like to learn a control strategy without knowing the desired output of the system at each time step in advance. Consider a sequence of actions on the cart, followed by a failure signal that means the pole has fallen. Let a 1 indicate a push to the right and a 0 to the left; "F" indicates failure. Consider the following sequence: 100011110000F. A classic credit assignment problem exists: which actions contributed to success and which actions contributed to failure?

At any time, the available state information includes the angle of the pole, θ, and the angular velocity of the pole, $\dot{\theta}$, as well as the position of the cart, ρ, and the velocity of the cart, $\dot{\rho}$. Using θ, $\dot{\theta}$, ρ, and $\dot{\rho}$ as inputs, the problem is to learn a decision policy for applying one of two actions to the cart at each time step: full-push left or full-push right, as in bang-bang control. The system is simulated by numerically approximating the equations of motion using Euler's method with a time step of $\tau = 0.02$ seconds and discrete time equations of the form $\theta(t + 1) = \theta(t) + \tau\,\dot{\theta}(t)$. The sampling rate of the system's state variables is the same as the rate of application of the control force, which is equal to 50 Hertz.

The equations of motion for the above system are as follows:

$$\ddot{\theta}_t = \frac{mg\ sin\ \theta_t - cos\ \theta_t[F_t + m_p l\dot{\theta}_t^2\ sin\ \theta_t]}{(4/3)ml - m_p lcos^2\ \theta_t},\ \ddot{\rho}_t = \frac{F_t + m_p l[\dot{\theta}_t^2\ sin\ \theta_t - \ddot{\theta}_t\ cos\ \theta_t]}{m},$$

where

ρ is the cart position
$\dot{\rho}$ is the cart velocity
θ is the pole angle
$\dot{\theta}$ is the angular velocity of the pole
l is the length of the pole = 0.5 m
m_p is the mass of the pole = 0.1 kg
m is the mass of the cart and pole = 1.1 kg
F is the control force = ± 10 N
g is the acceleration due to gravity = 9.8m/sec^2

A failure signal is associated with θ falling past a particular angle or with the cart running into the ends of the track at ± 2.4 meters from center. The control problem is approximately linear for small values of θ, which is typically limited to a 12-degree range. Although the 12-degree limit is commonly used to generate a failure signal, some of our experiments attempt to balance the pole over a much larger range of angles and allows starting states with θ ranging up to ± 74 degrees.

4. Genetic reinforcement learning

Figure 2 illustrates the basic components of our genetic reinforcement learning paradigm. A real-valued string determines the weights in a neural network of predefined size and connectivity. The network is then applied to the reinforcement learning problem. The "system" could refer to either a real or simulated model of the task to be performed. In the pole-balancing problem, each real-value string in the population is decoded to form a network with five input units, five hidden units, and one output unit. The network is fully connected between the input layer and the hidden layer; the input layer is also fully and directly connected to the output unit. All five hidden units also feed into the output unit. This network configuration is the same as that used by Anderson (1989) with the AHC algorithm, thus facilitating comparisons. Since there are 35 links in the network, each string used by the genetic search includes (35 + 1) real values concatenated together.[1] Before any input is applied to the network, the four state variables are normalized between 0 and 1. A bias unit fixed at 0.5 is also used as a fifth input to the net; a weight from the bias unit to a hidden node (or output node) in effect changes the threshold behavior of that node. The action of the neural network for a particular set of inputs is determined from the activation of the output unit.

A random start state is supplied to the network, which determines the action on the cart. The new state of the system is then calculated and reintroduced as a new input to the net.

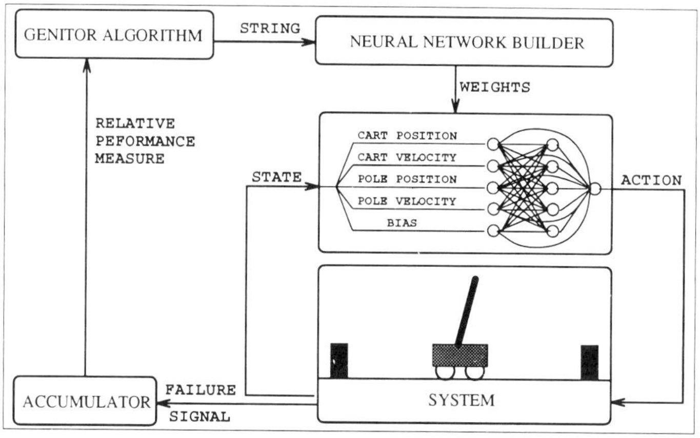

Figure 2. The genetic reinforcement learning paradigm.

This continues until a failure occurs. Feedback takes the form of an accumulator that determines how long the pole stays up and how long the cart avoids the end of the track; this is used as a relative measure of fitness.

Based on Anderson's earlier work (1989), we stopped learning when a network was found that was able to maintain the system without generating a failure signal for 120,000 time steps (40 minutes of simulated time). One potential problem with such a simple evaluation criterion is that a favorable or unfavorable start state may bias the fitness ranking of an individual net. In other words, the evaluation function is noisy. We would like to assign a fitness value to a string based on its ability to perform across all possible start states. In reality, this is not practical. In our initial experiments we started from one start state and interpreted the output action deterministically; the networks learned, but what was learned was not a full solution to the problem over all possible states. Networks starting from a favorable position may easily learn to keep the pole in position, but may not learn to deal with unfavorable conditions. We adopted the idea of interpreting the output action probabilistically from the AHC algorithms (Anderson, 1987) and found that this improved the performance of the controller. In other words, an output of +0.75 did not automatically mean that the action should be to push right, but rather that the probability of pushing right should be 0.75. This interpretation of the output action allows the network to visit more of the state space and hence to learn about more of the problem space.

The problem is to learn a total mapping policy from a limited sample of input training data that may be relatively uninformative. For genetic algorithms, this generalization problem is also a noisy "fitness" evaluation problem. Some sets of initial state variables guarantee failure for any sequence of control actions. Networks that receive poor starting states may be ranked lower than networks that receive good starting positions; some networks that represent good control policies will be lost. Methods to deal with this noise in the fitness evaluation function remain a topic of interest and are discussed later in this article. Fitzpatrick and Grefenstette (1988) show that the overall efficiency of a genetic algorithm may be improved by reducing the time spent evaluating each individual in favor of evaluating more individuals from a larger population. As will be shown in later sections, not all of our methods or results are directly comparable to Fitzpatrick and Grefenstette's work; however our results do support the notion that increasing population size can be a robust mechanism to obtain better generalization.

4.1. Initial results using genetic reinforcement learning

Initial experiments were carried out with a failure signal occurring when the pole angle falls past 12 degrees or the cart hits the end of the track. In these experiments the "fitness" of a string is based on the accumulated number of steps until failure, starting from a single random initial state vector. The genetic algorithm did not have any difficulty with this problem: the system learned to balance the pole on every attempt, as shown in table 1.

Experiments with supervised learning problems indicate that the genetic algorithm's tendency is to converge more quickly but less reliably using smaller populations (Whitley et al., 1991). A similar trend exists for the inverted pendulum problem. While the genetic algorithm learns in every case, smaller populations tend toward faster learning (as indicated by the median), but with poorer worst-case behavior. This sometimes skews the mean and results in a higher variance. Data supporting these observations are given in table 1 for population sizes (PopSize) 5, 50, and 100.

We also ran a second set of 50 experiments to determine if we could improve the learning speed of the genetic algorithm using some form of population initialization. Grefenstette (1987) discusses an initialization strategy where the population is seeded with heuristically improved strings. Our seeding strategy involved loading the initial population with strings that have already demonstrated the ability to balance the pole for 100 time steps. This was done by randomly generating strings until enough strings satisfying the entry requirement are obtained. This strategy helped to eliminate extreme worst-case behavior for small populations. However, since seeding the initial population results in additional computation for most individual runs of the genetic algorithm, it uniformly increases the median learning time across experiments using different population sizes. The experimental results in tables 1 and 2 for populations of 5, 50, and 100 strings suggest that for this problem the genetic hill-climber is fairly robust without seeding. In subsequent tests we used the genetic algorithm with a random initial population.

Table 1. "Learning rates" as measured by Best, Median, Worst, and Mean refer to the number of "starts" (evaluations) required to reach the first occurrence of a net that is able to balance the pendulum for 120,000 time steps. SD is the standard deviation of the number of starts.

Genetic learning rates using random initial populations Pole angle ± 12 degrees — Sample size: 50					
PopSize	Best	Worst	Median	Mean	SD
5	242	50255	1845	4544	9650
50	478	6308	2580	2855	1238
100	886	11481	3579	4097	2205

Table 2. The effect of seeding on learning rate for the genetic algorithm using population sizes (PopSize) of 5, 50, and 100. The figures refer to the total number of strings evaluated, including those evaluated while seeding the population.

Genetic learning rates for seeding experiments Pole angle ±12 degrees Sample size: 50					
PopSize	Best	Worst	Median	Mean	SD
5	880	14,287	2657	3491	2752
50	7357	16,915	11,427	11,512	1896
100	14,592	25,655	20,321	20,244	2399

4.2. Other genetic approaches to control problems

The use of genetic algorithms for control applications have been studied by several researchers. In reviewing this work we have limited our attention to three kinds of approaches. First are approaches that use a genetic algorithm to directly associate actions with all possible states in a discretized mapping of the state space. Second are approaches where genetic algorithms are used to evolve control policies in the form of rules. Third are approaches using genetic algorithms to train neural networks for control application.

Odetayo and McGregor (1989) as well as Thierens and Vercauteren (1990) have applied genetic algorithms to the pole-balancing problem by discretizing the space in a way that is similar to earlier approaches used by Michie and Chambers (1968) and Barto et al. (1983). When genetic search is applied to this discretized problem, it is used to find an appropriate action for each state-space partition. The genetic encoding is merely a binary string, where each bit represents an action (push left or right) for each partition of the space. One notable point is that the entire binary vector does not have to be correct for the entire space to keep the pole balanced; the pole need only be started in a favorable position and kept in a favorable position. However, this represents a failure to generalize the control strategy to all portions of the input space. Also, for problems that have a large number of variables, it becomes infeasible to discretize the space. Consider a problem with 30 variables; even if each variable is binary, the discretized space of 2^{30} partitions is unreasonable.

The work of Grefenstette and collaborators (Grefenstette, Ramsey, & Schultz, 1990; Grefenstette, 1991) on SAMUEL (Strategy Acquistion Method Using Empirical Learning) is relevant to the work presented here for several reasons. First, SAMUEL is explicitly designed to develop a strategy for sequential decision-making. One problem to which SAMUEL has been applied is the *Evasive Maneuvers* game, involving an agent trying to avoid an approaching predator. Second, while SAMUEL uses a symbolic rule-based approach to learning rather than a neural network, it still uses a genetic algorithm as the main learning component of the system. SAMUEL and genetic reinforcement learning therefore face many of the same issues related to learning: how to initialize the population, the use of different selection strategies, the choice of genetic operators (crossover and mutation) and—probably most important—the problem of noisy fitness evaluation. Noisy fitness evaluation is very relevant to control problems because we are using the genetic algorithm to evaluate the control strategy, represented by an encoded string, as to its ability to function for all possible inputs, while only actually testing the string on a small sample of cases. Whether the string is converted into a set of symbolic rules or connectionist weights is a secondary issue. While we have not compared our system to SAMUEL, such a comparison would be valuable.

More comparable to our own work are experiments carried out by Weiland (1990, 1991), who uses a standard generational genetic algorithm with a binary encoding of weights to train a fully recurrent network for several versions of the pole-balancing problem. Among these more difficult versions is a problem in which two poles are attached to the cart and simultaneously balanced; another problem involves balancing a jointed pole (i.e., one pole is attached to the top of another). These experiments also used the value of the output unit to determine the magnitude of the force applied to the cart.

The binary encoded genetic algorithms used by Weiland as well as the mapping of functionality to the recurrent network—and even the scaling of the binary encodings to the weights—are different enough that it is hard to compare Wieland's results with our own work. However, the limited success reported by Whitley, et al. (1990c) using binary encodings for the optimization of neural networks needs to be carefully reexamined in light of Weiland's results. We highly recommend Weiland's work to the reader interested in applying genetic algorithms to control problems.

5. AHC: Adaptive heuristic critic

In its search for a good network, the genetic reinforcement learning algorithm evaluates each candidate network by applying it to the balancing problem and measuring the time until failure. A new network is constructed after each balancing attempt. An alternative is to make small adjustments to one network after every step in a balancing attempt. The AHC algorithm (Barto et al., 1983; Sutton, 1988) takes this approach. Adjustments following each step are guided by information provided by a second network. This new network is called the evaluation network, while the original network that generates actions is called the action network. Figure 3 is a diagram of the evaluation and action networks. The action network has the same structure as the network in the genetic reinforcement learning model of figure 2. The evaluation network learns a function whose value indicates how often and how soon the inverted pendulum has failed after being in states similar to its current state. Thus, the AHC provides an evaluation of every state. A single action during a balancing attempt is rewarded or punished by comparing the evaluations of the state preceding an action and the state following the action. The difference between these evaluations is called a temporal-difference, or TD, error. Sutton (1988) defines a general class of TD methods for learning predictions. The AHC algorithm we used for our experiments is a member of this class. It is a general solution to the problem of distributing credit among the actions of a sequence that precedes a reinforcement. Sutton's analysis of linear temporal methods indicates that they are capable of learning a Markov model of the underlying system process. In the linear case there is a single "prediction-weight" associated with each state, and the output is a linear combination of the prediction-weight vector multiplied by the state vector. The result corresponds to the expected payoff associated with different absorbing states in the Markov model, where the payoff is adjusted to reflect the probabilities of entering

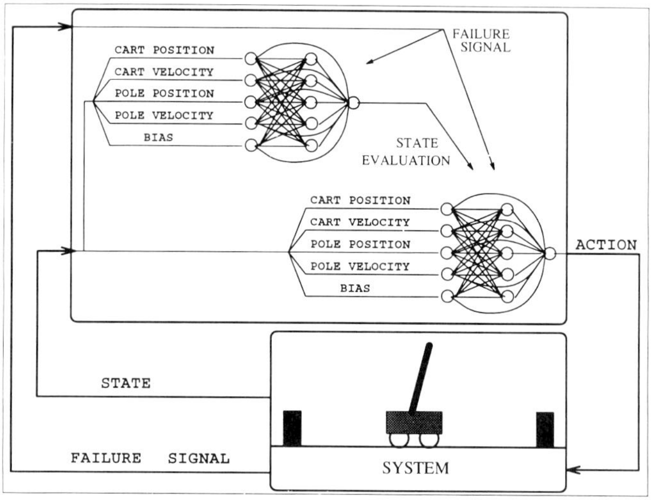

Figure 3. The adaptive heuristic critic learning paradigm.

the different absorbing states. Extending this to the nonlinear problem of predicting failure in the inverted pendulum problem, we would like the predictions associated with different state configurations to reflect the expected probability of making a sequence of moves that lead to failure, where a failure state is a terminal state with a payoff of -1 and all other states are nonterminal states with no payoff value.

After learning an approximately correct prediction of the failure signals, the evaluation network produces values between 0 and -1, with lower values indicating states from which failure is more likely. Barto et al. (1983) demonstated the use of the AHC for the pole-balancing problem when the cart-pole states are discretized. For the problem studied here without discretized states, reinforcement prediction as a function of cart-pole states is nonlinear. A multilayer AHC evaluation network with five hidden units in one hidden layer successfully learned this function. The hidden units of the evaluation network were trained using a modified error-propagation scheme where the error was a temporal-difference error. The AHC algorithm as described in Anderson (1987) is outlined in figure 4.

The action network has the same structure as the evaluation network, with five hidden units in one hidden layer. The hidden units of the action network are trained using a modified form of the back propagation algorithm (Anderson, 1987). Instead of having a predefined target output, the evaluation network is used to obtain evaluations of the current state variables. The difference between consecutive predictions in the evaluation network serves as a reinforcement of individual actions in the action network at each time step. The observed output of the action network is a continuous value giving the probability of pushing to the right. An action is chosen based on this probability. The difference between the actual action and the observed output is analogous to the difference between the target output and observed output terms normally used in back propagation. If the output action results in a state transition that reduces the prediction of failure, then the temporal difference error will be positive and the action will be positively reinforced. If the output action results

**** Initialization Phase ****

(1) Set each weight in the action network and the evaluation network to a random number from a uniform distribution from -0.1 to 0.1.

**** Iteration Phase ****

(2) Repeat until number of trials equals max-trials or number of steps equals max-steps:

(a) Forward pass through the evaluation network, calculating an evaluation, v_t, for the current state.
(b) Forward pass through the action network, calculating the probability, p_t, of pushing right. Using this probability an action is chosen, a_t, with value 0 or 1, representing pushes to the left or right, respectively.
(c) Apply action a_t to the pole simulation producing new pole state.
(d) Forward pass through evaluation network with new pole state, calculating an evaluation, v_{t+1}, for the new state.
(e) Calculate temporal difference between evaluations of new state and previous state. The temporal-difference error is $\gamma v_{t+1} - v_t$, where γ is a discount factor set to 0.9 in our experiments (c.f., Sutton 1988). If pole exceeds its bounds, $\gamma = 1$ and $v_{t+1} = -1$.
(f) Update the evaluation network's weights by back propagating the temporal-difference error. Normal back propagation calculates hidden nodes errors using the weights between the hidden nodes and the nodes they feed. This same calculation is applied, except only the signs of the weight value (± 1) are used.
(g) Update action network's weights by back propagating the product of the temporal-difference error and an action-comparison term given by $a_t - p_t$.
(h) If failure occurs, reset the pole state to random state and reset step counter to zero.

(3) Report results as the number of trials before a trial of max-steps has occurred.

Figure 4. The adaptive critic heuristic algorithm.

in a state transition that reduces the prediction of failure, then the temporal difference error will be positive and the action will be positively reinforced. If the output action results in a state transition that increases the prediction of failure, then the temporal difference error will be negative and the action is negatively reinforced. (For an alternative approach to reinforcement back propagation, see Ackley and Littman (1990).

The term "Adaptive Heuristic Critic" (AHC) is sometimes used in a more restrictive sense to refer only to the evaluation network that serves as a predictor of failure. However, the term has also been used in a more general sense to refer to the combined evaluation network and action network (e.g., Sutton, 1988, 1991). References to the "AHC algorithm" in this article also refer to the combined evaluation and action network when making comparisons to the genetic algorithm.

6. Comparative tests

We have attempted to compare genetic reinforcement learning to the adaptive-critic AHC architecture as used by Anderson (1987). It would be short-sighted to view these tests as merely a way of deciding if one algorithm is better than another. Our goal is to study the behavior of the learned action networks under different circumstances and to discover the respective strengths and weaknesses in both algorithms. The algorithms are compared at two levels. First, the consistency, speed, and reliability of learning itself is evaluated. Second, once the networks have been trained, how well do they generalize? Put another way, how can we characterize the performance of the networks across the application domain after training is complete? Besides offering comparisons of learning rates and performance, we also look at ways of improving performance.

6.1. Comparative test 1: Learning at 12, 35, and 74 degrees

The experiments of most researchers only attempt to learn to balance the pole within the 12-degree range. Additional experiments were carried out with failure signals occurring at three different positions of the pole: 12 degrees, 35 degrees, and 74 degrees. In the implementation used here, changing the angle at which the failure signal occurs also changes the range of the input variable representing the pole angle. We are therefore learning using a different representation of state space. The 12-degree restriction means that the inverted pendulum problem has a solution that is approximately linear. At 35 degrees the problem is nonlinear and contains many start states where it is impossible to balance the pole; empirically, we have found it is possible to balance the pole for states up to 39 degrees from vertical, but only if other states variables are favorable. Our empirical tests suggests that the pole will always fall when the pole angle is beyond 45 degrees, even if all state variables are favorable. Thus, at 74 degrees the space is dominated by start states from which it is impossible to balance the pole.

One fact that emerged from these tests is that AHC sometimes fails to learn. The genetic algorithm converged to a solution in every experiment, regardless of whether the failure signal occurred at 12, 35, or 74 degrees. Learning rates (the number of trials required for learning) for tests with failure signal set to 12 and 35 degrees are given in table 3. AHC learned a control strategy in 33 out of 50 attempts (66%) on the 12-degree problem. We also ran experiments using both algorithms with the failure signal extended to 35 and 74 degrees to increase the difficulty of the problem. AHC successfuly converged in 43 out of 50 attempts for the 35-degree problem. Both 12- and 35-degree experiments were run out to a maximum of 45,000 trials. Additional batteries of tests composed of 50 runs for both 12- and 35-degree problems showed that the convergence rate continued to fluctuate

Table 3. Comparison of learning rates for the AHC algorithm and the genetic algorithm, GA-100 (population size 100), using 12- and 35-degree failure signals. Results are averaged over 50 experiments for the genetic algorithm, and only over those cases that converged for the AHC.

	Learning Rates at 12 degrees					Learning Rates at 35 degrees				
Method	Best	Worst	Median	Mean	SD	Best	Worst	Median	Mean	SD
AHC	3182	13543	4529	5433	2390	2106	12076	2758	3922	2452
GA-100	886	11481	3579	4097	2205	820	7221	4231	4206	1777

between 66% and 86%, but that the convergence rate is independent of the bounds on the pole angle. In all of our experiments, if the AHC algorithm failed to learn by 15,000 trials, it never converged no matter what version of the problem was being learned.[2]

When the problem was extended to use a failure signal at 74 degrees, the AHC algorithm converged to a successful solution in only 1 out of 50 attempts (learning time, 8057 trials) using up to 100,000 learning trials. On the 74-degree problem, the learning rates of the genetic algorithm using a population of 100 are as follows: best: 6055, worst: 37,032, median: 17,972, mean: 18,921, SD: 7355.

The differences in the learning rates reported here are not statistically significant, but the genetic algorithm is clearly competitive with the AHC algorithm. Because the AHC does not reliably converge to a solution, we conclude that the genetic algorithm is more robust with respect to the rate of convergence.

6.2. Comparative test 2: Performance-based generalization

The most straightforward test of performance is a test of the networks trained by the AHC algorithm and by genetic reinforcement learning across both random and fixed test points drawn as samples from the entire state space. Initially each system was tested for different "fixed" initial state variables, with each of the four normalized state variables having the following values: 0.05, 0.275, 0.50, 0.725, 0.95. This resulted in 625 different initial positions. We also compared the networks on 625 randomly chosen start positions and found similar results. All subsequent comparisons used the 625 randomly chosen positions. All initial states are chosen by assigning random values to the state variables within the normalized input range; thus, the initial pole angle is within the 12-degree (or 35-degree) bounds. It might come as a surprise to find that so many of the starting positions result in a failure even for those networks that display the best performance. It would appear that there are many initial positions that are irrecoverable for any control strategy. Not all AHC action networks were tested; rather, networks were tested only for those cases where the learning criteria was successfully met.

Table 4 provides statistics about the performance of the neural networks found by each method. Each network is tested starting at 625 different random initial states; we then count in how many of the 625 tests each neural network is able to balance the pole for 1000 time

Table 4. Performance is measured by the number of start states for which each network successfully balances the pole for 1000 time steps when tested over 625 random start states. Results are shown for both the 12- and 35-degree problems. The genetic algorithm using a population of 100 (GA-100) is compared to the AHC algorithm. Only those networks that converged for the AHC algorithm were used in these comparisons.

	Performance Metrics for 625 Starting Positions									
	12-deg. Failure Signal					35-deg. Failure Signal				
Method	Best	Worst	Median	Mean	SD	Best	Worst	Median	Mean	SD
AHC	372	70	196	192	79	406	19	296	271	108
GA-100	446	24	315	297	89	430	92	337	304	92

Table 5. Performance is measured by the number of start states for which each network successfully balances the pole for 1000 time steps when tested over 625 random start states. This table shows the effect on performance of varying the population size (PopSize) used by the genetic algorithm for the 12-degree problem.

PopSize vs. Performance at 12 degrees					
PopSize	Best	Worst	Median	Mean	SD
5	424	78	266	262	81
50	411	63	281	253	102
100	446	24	315	297	89

steps. This is enough time to either recover or fail in a difficult situation. The best networks are able to balance the pole from more than 400 of the 625 initial start states.

The same failure criteria used for training were also used for testing. Several things are interesting about the performance results. First, the data indicate that performance generally increased for the genetic algorithm using the larger population of 100 strings (see table 5). A population of 100 found the best overall net, as well as the most desirable mean and median case behavior. In all of our tests the genetic algorithm produced results than are competitive with the AHC algorithm, in terms of learning rates, convergence behavior, and performance.

6.3. Comparative test 3; Tracking control behavior over time

We developed several ways of tracking the control behavior of these networks over time. Obviously, one can animate the cart and pole system—and we did this. While animation was useful, we found that graphs that mapped the input and output behavior of the networks over time were much more informative. Figures 5 and 6 are made up of two different subgraphs. The top subgraph tracks the four *normalized* state variables (which serve as the neural network inputs) as a function of time. The bottom subgraph tracks the activation of the output unit, which is continuous. During training, the output is determined probabilistically depending on the activation of the output unit. During testing, the action applied to the system is obtained by deterministically thresholding the activation value of the output unit. If the activation value is greater than 0.5, then output a 1 and push right; if it is less than or equal to 0.5, then output a 0 and push left.

If the pole is vertical and the cart is centered and the velocities are 0, then all state variables will have the normalized value 0.5. When the system is started in an "ideal" state (all state variables are initialized to 0.5), then a successfully trained network will maintain the state variables close to the 0.5 level. It is not possible to balance the pole and avoid the track terminals from all possible initial states; however, when started from an initial state from which recovery is possible, a perfectly trained network should drive all state variables back to the 0.5 level representing the ideal state of the system.

In figures 5 and 6, the cart is at the far right end of the track with the pole leaning 32 degrees to the left; the 12-degree failure signal is disabled for these tests. The cart velocity and pole velocity are initialized at 0 (i.e., 0.5 is the normalized form). This initial state

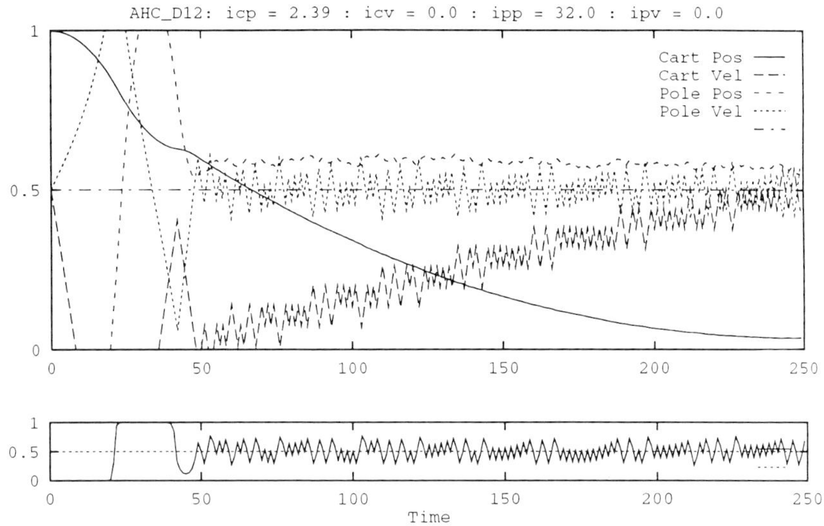

Figure 5. Control behavior of the best AHC-trained network for the 12-degree problem. The top subgraph tracks the four normalized state variables as a function of time. The bottom subgraph tracks the activation of the output unit, which determines the direction the cart is pushed. The cart is initially positioned at the far right end of the track with the pole leaning 32 degrees to the left. Since the 32-degree starting angle of the pole exceeds 12 degrees, the plot of the pole angle is initially beyond the range used in this graph. The initial cart and pole velocities are both set to zero.

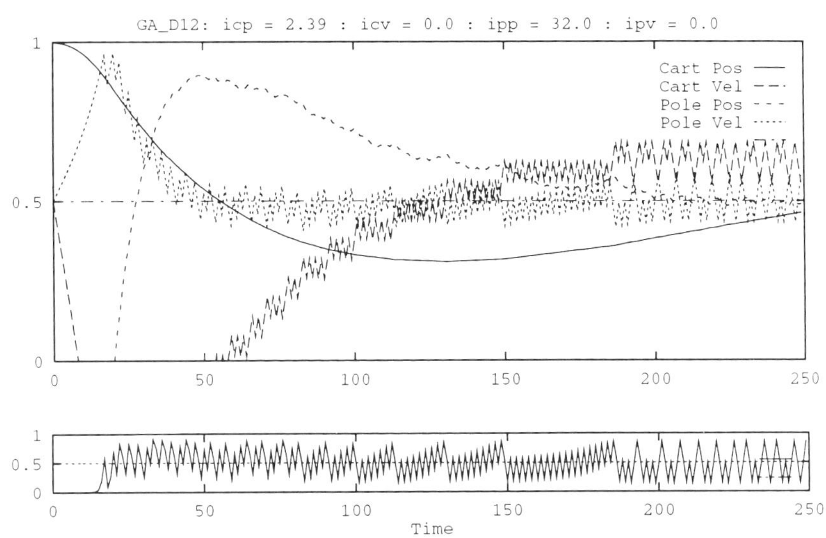

Figure 6. Control behavior of the best genetically trained network for the 12-degree problem. The cart is initially positioned at the far right end of the track with the pole leaning 32 degrees to the left.

constitutes a position from which it is difficult for the system to recover. Both the AHC-trained network and the genetically trained network used to produce these graphs are the best networks obtained for the 12-degree problem, as indicated by the performance tests summarized in table 4. Figure 6 shows that the genetically trained network gets all of the input variables into tolerable ranges fairly quickly, whereas the AHC-trained network (figure 5) takes longer. The AHC-trained network quickly dampens pole velocity and reduces oscillation in the pole position, but in doing so the cart almost crashes into the opposite end of the track. The genetically trained network handles problems with starting pole angles beyond 32 degrees, but the AHC-trained network does not. Results from the next section help to explain this behavior.

Figure 7 and figure 8 show results for an AHC-trained network and a genetically trained network using a failure signal at 35 degrees during learning. Again we use the best networks according to the performance data reported in table 4. These graphs indicate that both the AHC-trained network and the genetically trained network exploit similar information to determine the output activation levels and that they employ similar control strategies. The networks trained at 35 degrees proved to be more similar across a wider range of start states, but as the difficulty of the initial start states is increased the AHC-trained networks fail sooner than the genetically trained networks. In these graphs, the system is started with the cart in the same far right position and the pole leaning 34 degrees to the left. Cart velocity and pole velocity are initially 0.

These graphs make it evident that both networks track pole velocity by varying the magnitude of the output value. The correlation between pole velocity and the output activation is not as discernible in the first 50 to 100 time steps because the system is recovering from a difficult initial situation; correlation between the pole velocity and the output activation

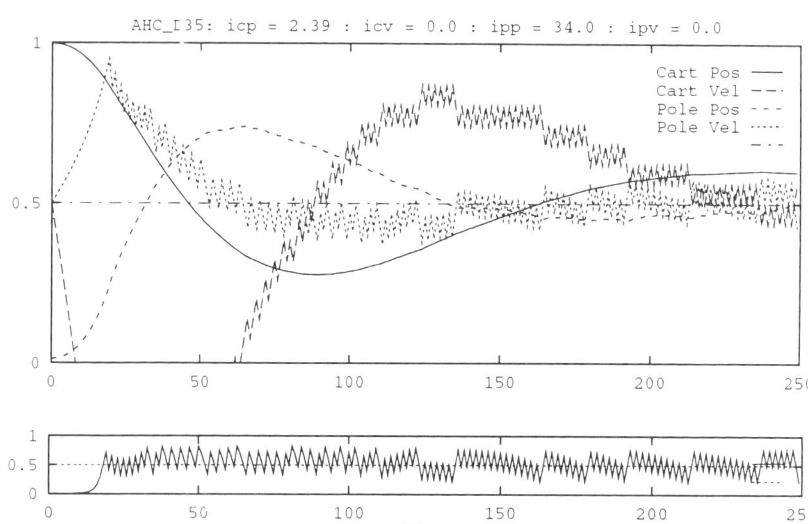

Figure 7. Control behavior of the best AHC-trained network for the 35-degree problem. The cart is initially positioned at the far right end of the track with the pole leaning 34 degrees to the left.

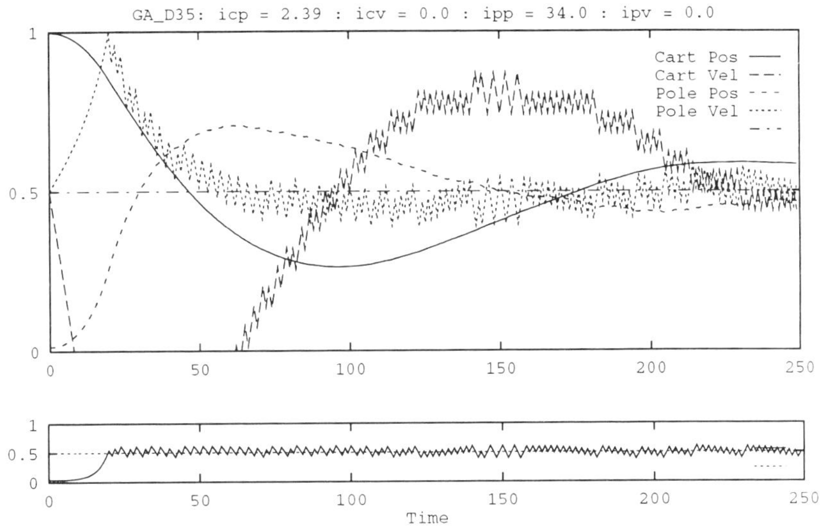

Figure 8. Control behavior of the best genetically trained network for the 35-degree problem. The cart is initially positioned at the far right end of the track with the pole leaning 34 degrees to the left.

is much more pronounced as the networks begin to bring the system under control. Also notable is that cart velocity and pole velocity tend to be negatively correlated. Given the input definitions used in our experiments, cart velocity and pole velocity have a similar, but opposite relationship. In the next section we prune the neural networks in order to isolate possible functional relationships between the state variables and the output unit's activation.

6.4. Comparative test 4: Pruning networks to find critical variables

To better understand the control strategies developed by the two algorithms, we attempted to reduce several successful networks to their minimal form. One way to prune a network is to remove any hidden unit whose activation is approximately constant for all inputs. Such a hidden unit can be eliminated by combining it with the bias unit (which always has a constant activation value). The weights that fan out from the eliminated unit are absorbed by appropriately modifying the weights that fan out from the bias unit. Another way to reduce the number of hidden units in the network is to combine two units with similar output behavior. A hidden unit p can be removed when its output is approximately equal to the output of another hidden unit q for all input pattern. For all units r, which receive input from both p and q, the value of the weight w_{pr} is added to the weight w_{qr}. The unit p can now be eliminated along with all weights associated with it. Sietsma and Dow (1991) give a more detailed explanation of these kinds of methods that allow a user to hand-prune a neural network.

The networks that were reduced are the best networks from table 4 for a failure signal of both 12 degrees and 35 degrees. In each case we were able to reduce the number of

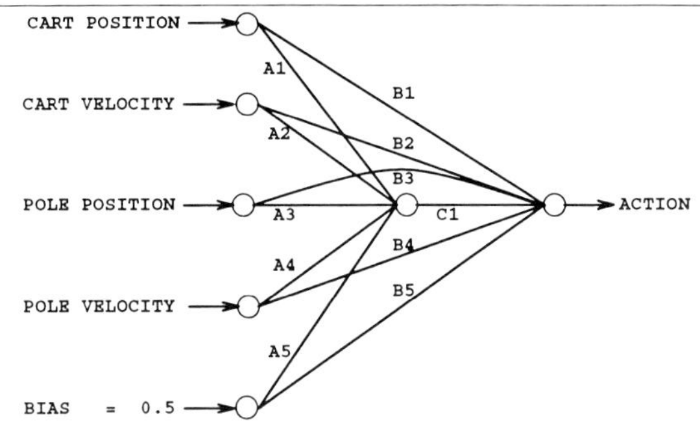

NET	WEIGHT										
	A1	A2	A3	A4	A5	B1	B2	B3	B4	B5	C1
AHC-12	-0.5	-0.5	-0.5	-0.5	0.0	0.0	0.0	16.0	16.0	-8.0	-45.0
GA-12	14.2	-1.3	16.4	11.2	-1.4	3.71	7.72	14.2	33.5	-12.6	-93.6
AHC-35	-0.5	-0.5	-0.5	-0.5	0.0	0.93	0.44	12.5	20.3	-8.3	-44.3
GA-35	-0.39	-2.49	-3.77	-2.49	7.82	4.84	0.38	7.85	6.45	-16.0	-4.37

Figure 9. The pruned forms of the neural networks produced by the AHC and genetic algorithms.

hidden units down to one; this pruning still allowed performance to remain within $\pm 1\%$ of performance before pruning as determined using the 625 fixed test points.

Figure 9 shows the reduced networks. An examination of numerous networks trained by the AHC algorithm reveals that the action networks always have a similar weight pattern. Weights leading from the input layer to the hidden layer are always close to the -0.5 range (see connections A1–A4). Larger weights are typically found on connections that feed into the output, particularly the direct input/output connections (e.g., B3–B5). This suggests that the computation of the AHC networks may be dominated by these linear associations. The weights for the genetically trained networks were more difficult to analyze and prune.

The weights in figure 9 suggest that the genetically trained network for the 12-degree problem uses more information about cart velocity and position than the AHC trained network (where B1 and B2 are 0). To further test this hypothesis, we started a cart on the far right of the track with the pole angle at 36 degree to the left and a high cart velocity to the left. We also reset the failure signal to occur at ± 90 degrees. The networks trained at 12 degrees were than tested. The genetically trained network got the pole up using the favorable velocity, then got the velocity under control, thereby recovering from a difficult situation. The AHC-trained network was able to get the pole up, but failed to control the velocity of the cart, which crashed into the left track terminal. None of the AHC networks trained at 12 degrees could recover if the pole angle fell past 32 degrees.

The graphs in section 6.3 suggest that the computation of the genetic trained networks and the AHC-trained networks are similar, but also indicate that the genetically trained networks are better able to recover from extreme initial states. The weights obtained by pruning the neural networks suggest that this difference may be due to better exploitation of state information about cart position and velocity.

6.5. Comparative test 5: A tougher learning criteria

Both the genetic and AHC algorithms show a good deal of variability in what is learned from one experiment to the next. This is perhaps not surprising, considering the stopping criteria. Controlling the inverted pendulum so that it successfully balances for 120,000 time steps from a single random starting position does not represent a consistent metric of success from one experiment to the next because the ability to balance for 120,000 time steps is influenced by the initial state vector. There is some evidence to suggest that action networks from the experiments with the fastest learning times have poorer performance. Sammut and Cribb (1990) conjecture that in general that there is a tradeoff between learning rate and the ability to generalize. Specifically, they found experimentally that programs that very rapidly learned to perform a task resulted in very specific control strategies that could not be transferred to other initial start states.

Since the fastest learning times may be correlated with fortuitious starting positions, a stopping criteria that should be more indicative of overall performance was chosen. Both the AHC and genetic algorithms were stopped only after demonstrating success at the learning task multiple times. The first increment in the stopping criteria was to do two consecutive balancing attempts successfully for 120,000 steps (the "2/2 rule"). Each additional attempt to balance the pole is carried out from a new random initial start state. This appears to result in improved performance behavior for the genetic algorithm, but no clear trend emerged for the AHC algorithm. The stopping criteria was further extended to three and four consecutive balances (the "3/3 rule" and the "4/4 rule"). As can be seen in table 6, an additional improvement in mean performance and reduced variance is again observed for the genetic algorithm.

Using a more difficult stopping criteria resulted in modest increases in training time. But the data in table 6 makes it clear that simple comparisons of learning speed can sometimes be very misleading if the effect on performance is not also considered. Using the genetic algorithm, the mean performance of the resulting networks improved and the performance variance decreased as the stopping criteria became more challenging. One other notable point, which we have not explored, is the potential for further improving the performance

Table 6. Results of using a stricter stopping criteria. The genetic algorithm (GA-100) uses a population size of 100.

Using Different Stopping Criteria at 12 degrees								
	Learning Rates				Performance Metrics			
Method	Best	Mean	Median	SD	Best	Mean	Median	SD
AHC 1/1 rule	3182	4529	5433	2390	372	196	191	79
AHC 2/2 rule	3190	4628	5331	2437	377	183	193	88
AHC 3/3 rule	3600	4740	5204	1754	324	192	188	79
AHC 4/4 rule	3228	4702	5384	1763	369	175	180	72
GA-100 1/1 rule	886	3579	4097	2205	446	315	291	89
GA-100 2/2 rule	689	4760	5148	3323	425	328	307	82
GA-100 3/3 rule	232	4720	5226	2727	445	330	326	66
GA-100 4/4 rule	1526	5507	7071	7602	428	371	360	46

of the genetically trained networks by using a larger population size during learning. However, much larger population sizes could aggravate the *structural/functional mapping problem* discussed earlier in this article.

Convergence rates continued to fluctuate between 66% and 86% for the AHC using different stopping criteria. The criteria for stopping do not appear to be factors in the probability of convergence, however. We found *if an AHC network balanced the pendulum once, it always continued learning to a more difficult criterion.* We again note that all of the AHC-trained networks that we have developed either learn before 15,000 evaluations or entirely fail to learn. The fact that longer training times did not help with the convergence problem may also mean that it is not beneficial to lengthen training time by using a stricter stopping criterion.

7. Genetic algorithms and noisy fitness functions

As one might expect there appears to be a relationship between larger population sizes and more reliable control functions being learned by the genetic algorithm. Work by Fitzpatrick and Grefenstette (1988) on an image registration problem using a noisy evaluation function also indicates that fewer samples are needed to estimate the "correct" fitness of a string using a noisy evaluation when larger populations are used. This is because in the larger populations we expect more schemata samples; thus the effect of the noise on the fitness of each (implicitly estimated) schema is averaged out at the schema level rather than at the string level. This implies that a tradeoff exists between relying on a smaller number of more accurate (and costly) evaluations versus relying on a larger number of less accurate (and less costly) evaluations. Grefenstette et al. (1990) report similar findings in their application of genetic learning to the *evasive maneuvers* control problem.

When an evaluation function is noisy, it may be possible to average the performance from several initial state vectors to obtain a more accurate evaluation. We therefore thought that the performance of the genetic algorithm could perhaps be improved for the inverted pendulum problem by averaging fitness over three random start positions. The results were somewhat surprising. There was no improvement in performance, and the training time more than doubled (although the total number of "fitness evaluations" was reduced). The control policies of the resulting action networks were no more robust than controllers trained normally. One possible hypothesis that could explain our results relates to the interpretation of the output of the network as a probabilistic value. By interpreting the output of the network probabilistically, we presumably visit more of the state space, which is perhaps analogous to averaging over multiple starts using a deterministic interpretation of the output. Experiments to test this hypothesis could be designed and would provide valuable information. Overall, the use of the stricter stopping criteria was more effective at improving performance than averaging fitness over multiple tests of the neural net.

Several factors complicate the current experiments so that it is more difficult to consider the effects of noise on schema samples. First, it should be noted that the image registration work (Fitzpatrick & Grefenstette, 1988) was done in the context of a standard genetic algorithm using generational replacement; as recently noted by Davis (1991), the effects of noisy evaluation in the context of a one-at-a-time replacement scheme remains an open issue.

In a generational model, strings are evaluated each generation; if the evaluation function is noisy, a more "correct" estimate of the average value of strings in a particular hyperplane region can be obtained by using larger populations to average out the noise and by generational evaluation to reevaluate the strings. The use of the GENITOR algorithm with one-at-a-time reproduction and replacement could pose a potential problem, because once a string is evaluated, it is ranked in the population and never reevaluated. If the fitness function is noisy, as it is in this case, this means the ranking will include errors. However, in this particular application, noisy evaluation did not prevent the algorithm from learning. We believe this is because the noise largely had a conservative effect: some good networks are lost because of poor start states, but it is more difficult for a poor net to obtain a good ranking.

8. Discussion and conclusions

There are several differences in the information used by the AHC and genetic algorithms, at least in the learning component. In both cases, the action network requires input information about the current state in order to produce an output. For the AHC algorithm, the evaluation network also requires state information to learn, since without state information there is no prediction of failure. But the genetic algorithm does not require state information at each time step; it only needs feedback about how long the pole stayed up in order to rank competing sets of weights. The reinforcement back propagation that occurs in the AHC algorithm means that learning continues even when failures are not occurring. Furthermore, the use of the temporal difference method means that the evaluation network is also being updated, even when failures are not occurring. In the genetic approach used here, updates to the action network occur only after one or more failures: learning is not continuous. Another difference is that the genetic approach used in our experiments will assign an equal evaluation to two networks that avoid failure for an equal number of time steps. However, the AHC algorithm evaluates networks by the trajectory of states that are experienced. The evaluations associated with any two networks would differ when the AHC training algorithm is used, favoring the network that drives the cart-pole through more highly valued states. The restriction of the search to highly valued states may also explain why the performance of AHC trained networks did not improve when the stricter stopping criterion was used.

Yet another difference between the genetic algorithm and the AHC algorithm is the lack of success using the AHC algorithm when the failure signal occurs at wider pole bounds. This possibly results from the combination of the incremental learning algorithm used to adjust the weights and the way the AHC evaluation network generalizes. A good prediction of failure is hard to learn when the majority of start states lead to failure. Without a good failure prediction function, a successful control strategy cannot be learned using the reinforcement-driven back propagation. The genetic algorithm, because it ranks each network based on performance, is able to ignore those cases where the pole cannot be balanced; only the successful cases will obtain the chance to engage in genetic reproduction. For the AHC evaluation network, however, the preponderance of failures may cause all states to overpredict failure. This problem can be corrected either by selectively sampling the space to extract a better balance of success and failure information or by tuning the AHC algorithm to place more emphasis on positive results and less on failure.

Our purpose in the current study is (1) to demonstrate that genetic hill-climbing algorithms can be used for training neural networks for control problems, and (2) to examine methods for comparing the learning behavior and performance of algorithms for neurocontrol problems. The differences between genetic reinforcement learning and reinforcement learning using other methods must be explored more carefully, especially in other application domains. Anderson and Miller (1990) discuss several challenging control problems that could provide an initial test bed. We are also interested in ways in which these two different algorithms might be combined or hybridized. Recently, Ackley and Littman (1991) have combined genetic methods and neural networks for control problems in a different way. They use a genetic algorithm to train an evaluation net, then use the output of the evaluation net to do reinforcement error propagation on the action net. This represents a strategy that is intermediate between the methods compared in this article.

The results that we report here represent just one approach to genetically train neural networks for control problems. These are numerous questions that remain to be answered. There are other forms of adaptive critics, such as Q-learning algorithms (Watkins, 1989; Sutton, 1991), to which we could compare results. There is also a critical need to test different approaches across a suite of test problems that display varying levels of difficulty. Overall, we are very encouraged by the results obtained so far. The comparative methods we have used in this article also highlight the need for a broadly based comparative perspective when studying different reinforcement learning algorithms for control problems.

Acknowledgments

This research was supported in part by NSF grant IRI-9010546 and in part by a grant from the Colorado Institute of Artificial Intelligence (CIAI). CIAI is sponsored in part by the Colorado Advanced Technology Institute (CATI), an agency of the State of Colorado.

Notes

1. The extra real value encodes the crossover probability of the string.
2. In the final stages of preparing this article, we encountered one AHC-trained network for a 35-degree problem that converged after 39,000 trials; this was the only such occurrence out of several hundred tests.

References

Ackley, D., & Littman, M. (1990). Generalization and scaling in reinforcement learning. In D. Touretzky (Ed.), *Advances in neural information processing systems* (Vol. 2). San Mateo, CA: Morgan Kaufmann.

Ackley, D., & Littman, M. (1991) *Interactions between learning and evolution*. Morristown, NJ: Cognitive Science Research Group, Bellcore.

Anderson, C.W. (1987). *Strategy learning with multilayer connectionist representations* (TR87-509.3). GTE Labs, Waltham, MA.

Anderson, C.W. (1989). Learning to control an inverted pendulum using neural networks. *IEEE Control Systems Magazine*, *9*, 31–37.

Anderson, C.W., & Miller, W.T. (1990). A challenging set of control problems. In T. Miller, R. Sutton, & P. Werbos (Eds.), *Neural networks for control*. Cambridge, MA: MIT Press.

Barto, A.G. (1990). Connectionist learning for control. In T. Miller, R. Sutton, & P. Werbos (Eds.), *Neural networks for control*. Cambridge, MA: MIT Press.

Barto, A.G., Sutton, R.S., & Anderson, C.W. (1983). Neuronlike adaptive elements that can solve difficult learning control problems. *IEEE Transactions on Systems, Man, and Cybernetics*, *13*, 834–846.

Booker, L., Goldberg, D., & Holland, J. (1989). Classifier systems and genetic algorithms. *Artificial Intelligence*, *40(1-3)*, 235-282.

Davis, L. (1989). Adapting operator probabilities in genetic search. *Proceedings of the Third International Conference on Genetic Algorithms* (pp. 61–69). Fairfax, VA: Morgan Kaufmann.

Davis, L. (1991). *The handbook of genetic algorithms*. New York: Van Nostrand Reinhold.

Fahlman, S., & Lebiere, C. (1990). The cascade correlation learning architecture. In D. Touretzky (Ed.), *Advances in neural information processing systems* (Vol. 2). San Mateo, CA: Morgan Kaufmann.

Fitzpatrick, J., & Grefenstette, J. (1988). Genetic algorithms in noisy environments. *Machine Learning*, *3(2–3)*, 101–120.

Goldberg, D. (1989). *Genetic algorithms in search, optimization and machine learning*. Reading, MA: Addison-Wesley.

Grefenstette, J. (1987). Incorporating problem specific knowledge into genetic algorithms. In L. Davis (Ed.), *Genetic algorithms and simulated annealing*. London: Pitman/Morgan Kaufmann.

Grefenstette, J. (1989). A system for learning control strategies using genetic algorithms. *Proceedings of the Third International Conference on Genetic Algorithms* (p. 183–190). Fairfax, VA: Morgan Kaufmann.

Grefenstette, J. (1991). Strategy acquisition with genetic algorithms. In L. Davis (Ed.), *Handbook of Genetic Algorithms*. New York: Van Nostrand Reinhold.

Grefenstette, J.J., Ramsey, C.L., & Scultz, A.C. (1990). Learning sequential decision rules using simulation models and competition. *Machine Learning*, *5*, 355–381.

Harp, S., Samad, T., & Guha, A. (1990). Designing application-specific neural networks using the genetic algorithm. *Neural Information Processing Systems* (Vol. 2). San Mateo, CA: Morgan Kaufman.

Holland, J. (1975) *Adaptation in natural and artificial systems*. Ann Arbor, MI: University of Michigan Press.

Holland, J. (1986). Escaping brittleness: The possibilities of general purpose learning algorithms applied to parallel rule-basd systems. In R. Michalski, J. Carbonell, & T. Mitchell (Eds.), *Machine learning* (Vol. 2). San Mateo, CA: Morgan Kaufmann.

Michie, D., & Chambers, R. (1968). BOXES: An experiment in adaptive control. In E. Dale & D. Michie (Eds.), *Machine intelligence* (Vol. 2). Edinburgh: Oliver and Boyd.

Miller, G., Todd, P., & Hegde, S. (1989). Designing neural networks using genetic algorithms. *Proceedings of the Third International Conference on Genetic Algorithms* (pp. 379–384). Fairfax, VA: Morgan Kaufmann.

Montana, D., & Davis, L. (1989). Training feedforward neural networks using genetic algorithms. *Proceedings of the 1989 International Joint Conference on Artificial Intelligence* (pp. 762–767).

Odetayo, M.O., & McGregor, D.R. (1989). Genetic algorithm for inducing control rules for a dynamic system. *Proceedings of the Third International Conference on Genetic Algorithms* (pp. 177–182). Fairfax, VA: Morgan Kaufmann.

Sammut, C., & Cribb, J. (1990). Is learning rate a good performance criterion for learning. *Machine Learning: Proceedings of the 7th International Conference* (pp. 170–178). San Mateo, CA: Morgan Kaufmann.

Schaffer, D. (1987). Some effects of selection procedures on hyperplane sampling by genetic algorithms. In L. Davis (Ed.), *Genetic algorithms and simulated annealing*. London: Pitman/Morgan Kaufmann.

Schaffer, J.D., Caruana, R.A., & Eshelman, L.J. (1990). Using genetic search to exploit the emergent behavior of neural networks. *Physica D*, *42*, 244–248.

Selfridge, O.G., Sutton, R.S., & Barto, A.G. (1988). Training and tracking in robotics. *Proceedings of the Fifth International Conference on Machine Learning* (pp. 437–443). San Mateo, CA: Morgan Kaufmann.

Sietsma, J., & Dow, R. (1991). Creating artificial neural networks that generalize. *Neural Networks*, *4*, 67–79.

Sutton, R. (1988). Learning to predict by the methods of temporal differences. *Machine Learning*, *3*, 9–44.

Sutton, R. (1991). Reinforcement learning architectures for animats. In J. Meyers & S. Wilson (Eds.), *Simulation of adaptive behavior: From animals to animats* (pp. 288–296). Cambridge, MA: MIT Press.

Thierens, D., & Vercauteren, L. (1990). A topology exploiting genetic algorithms to control dynamical systems. In H.P. Schwefel & R. Manners (Eds.), *Parallel problems solving from nature* (pp. 104–108). Springer/Verlag.

Watkins, C. (1990). *Learning with delayed rewards*. Ph.D. dissertation, Psychology Department, Cambridge University, Cambridge, England.

Weiland, A. (1990). Evolving controllers for unstable systems. In D. Touretzky, J. Elman, T. Sejnowski & G. Hinton (Eds.), *Connectionist models: Proceedings of the 1990 Summer School* (p. 91–102). San Mateo, CA: Morgan Kaufmann

Weiland, A. (1991). Evolving neural network controllers for unstable systems. *1991 International Joint Conference on Neural Networks, 2* (pp. 667–673). Seattle.

Werbos, P. (1989). Backpropagation and neurocontrol: A review and prospectus. *1989 International Joint Conference on Neural Networks, 1* (pp. 209–215). Washington, DC.

Werbos, P. (1990). A menu of designs for reinforcement learning over time. In T. Miller, R. Sutton, & P. Werbos (Eds.), *Neural networks for control*. Cambridge, MA: MIT Press.

Whitley, D., & Kauth, K. (1988). GENITOR: A different genetic algorithm. *Proceedings of the 1988 Rocky Mountain Conference on Artificial Intelligence* (pp. 118–130). Denver.

Whitley, D., & Hanson, T. (1989). Optimizing neural nets using faster, more accurate genetic search. *Proceedings of the Third International Conference on Genetic Algorithms* (pp. 391–396). Fairfax, VA: Morgan Kaufmann.

Whitley, D., & Bogart, C. (1990a). The evolution of connectivity: Pruning neural networks using genetic algorithms. *1990 International Joint Conference on Neural Networks, 1* (p. 134–137). Washington, DC.

Whitley, D., & Starkweather, T. (1990b). Optimizing small neural networks using a distributed genetic algorithm. *1990 International Joint Conference on Neural Networks, 1* (pp. 206–209). Washington, DC.

Whitley, D., Starkweather, T., & Bogart, C. (1990c). Genetic algorithm and neural networks: Optimizing connections and connectivity. *Parallel Computing, 14*, 347–361.

Whitley, D., Dominic, S., & Das, R. (1991). Genetic reinforcement learning with multilayered neural networks. In R. Belew & L. Booker (Eds.), *Proceedings of the 4th International Conference on Genetic Algorithms* (pp. 562–569). San Diego, CA: Morgan Kaufmann.

Received November 21, 1991
Accepted March 24, 1992
Final Manuscript July 23, 1992

Machine Learning, 13, 285–319 (1993)

What Makes a Problem Hard for a Genetic Algorithm? Some Anomalous Results and Their Explanation

STEPHANIE FORREST FORREST@CS.UNM.EDU
Department of Computer Science, University of New Mexico, Albuquerque, NM 87181-1386

MELANIE MITCHELL* MM@SANTAFE.EDU
Artificial Intelligence Laboratory, University of Michigan, Ann Arbor, MI 48109-2110

Abstract. What makes a problem easy or hard for a genetic algorithm (GA)? This question has become increasingly important as people have tried to apply the GA to ever more diverse types of problems. Much previous work on this question has studied the relationship between GA performance and the structure of a given fitness function when it is expressed as a *Walsh polynomial*. The work of Bethke, Goldberg, and others has produced certain theoretical results about this relationship. In this article we review these theoretical results, and then discuss a number of seemingly anomalous experimental results reported by Tanese concerning the performance of the GA on a subclass of Walsh polynomials, some members of which were expected to be easy for the GA to optimize. Tanese found that the GA was poor at optimizing all functions in this subclass, that a partitioning of a single large population into a number of smaller independent populations seemed to improve performance, and that hillclimbing outperformed both the original and partitioned forms of the GA on these functions. These results seemed to contradict several commonly held expectations about GAs.

We begin by reviewing *schema processing* in GAs. We then give an informal description of how Walsh analysis and Bethke's Walsh-schema transform relate to GA performance, and we discuss the relevance of this analysis for GA applications in optimization and machine learning. We then describe Tanese's surprising results, examine them experimentally and theoretically, and propose and evaluate some explanations. These explanations lead to a more fundamental question about GAs: what are the features of problems that determine the likelihood of successful GA performance?

Keywords. Genetic algorithms, Walsh analysis, Tanese functions, deception.

1. Introduction

The genetic algorithm (GA) is a machine-learning technique, originated by Holland (1975), loosely based on the principles of genetic variation and natural selection. GAs have become increasingly popular in recent years as a method for solving complex search problems in a large number of different disciplines. The appeal of GAs comes from their simplicity and elegance as algorithms as well as from their power to discover good solutions rapidly for difficult high-dimensional problems. In addition, GAs are idealized computational models of evolution that are being used to study questions in evolutionary biology and population genetics (Bergman & Feldman, 1990).

In the simplest form of the GA, bit strings play the role of chromosomes, with individual bits playing the role of genes. An initial population of individuals (bit strings) is generated randomly, and each individual receives a numerical evaluation that is then used to make multiple copies of higher-fitness individuals and to eliminate lower-fitness individuals. Genetic

*Current address: Santa Fe Institute, 1660 Old Pecos Tr., Suite A, Santa Fe, NM 87501.

operators such as mutation (flipping individual bits) and crossover (exchanging substrings of two parents to obtain two offspring) are then applied probabilistically to the population to produce a new population, or *generation*, of individuals. The GA is considered to be successful if a population of highly fit individuals evolves as a result of iterating this procedure.

The GA has been used in many machine-learning contexts, such as evolving classification rules (e.g., Packard, 1990; De Jong & Spears, 1991), evolving neural networks (e.g., Miller, Todd, & Hegde, 1989; Whitley, Dominic, & Das, 1991), classifier systems (e.g., Holland, 1986; Smith, 1980), and automatic programming (e.g., Koza, 1990). In many of these cases there is no closed-form "fitness function"; the evaluation of each individual (or collection of individuals) is obtained by "running" it on the particular task being learned. The GA is considered to be successful if an individual, or collection of individuals, evolves that has satisfactorily learned the given task. The lack of a closed-form fitness function in these problems makes it difficult to study GA performance rigorously. Thus, much of the existing GA theory has been developed in the context of "function optimization," in which the GA is considered to be successful if it discovers a single bit string that represents a value yielding an optimum (or near optimum) of the given function. An introduction to GAs and GA theory is given by Goldberg (1989c), and many of the results concerning GA theory and applications can be found in the various ICGA proceedings (Grefenstette, 1985; Grefenstette, 1987; Schaffer, 1989; Belew & Booker, 1991), in Davis (1987, 1991), and in the two previous special issues of *Machine Learning* (Goldberg & Holland, 1988a; De Jong, 1990a).

GAs have been successfully applied to many difficult problems, but there have been some disappointing results as well. In cases both of success and failure, there is often little detailed understanding of why the GA succeeded or failed. Given the increasing interest in applying the GA to an ever wider range of problems, it is essential for the theory of GAs to be more completely developed so that we can better understand how the GA works and when it will be most likely to succeed.

In this article, we focus on a specific, initially surprising instance of GA failure that brings up several general issues related to this larger goal. We describe and explain a number of seemingly anomalous results obtained in a set of experiments performed by Tanese (1989a) using the GA to optimize a particular subset of Walsh polynomials. These polynomials were chosen because of their simplicity and the ease with which they can be varied, and because of the relationship between GAs and Walsh functions (see section 3). They were expected to present a spectrum of difficulty for the GA. However, the results of the GA on these polynomials were not at all as expected: for example, the GA's performance was strikingly poor on all of the functions included in Tanese's experiments, and was significantly worse than the performance of a simple iterated hillclimbing algorithm. Our analysis of these apparently surprising results explains the failure of the GA on these specific functions, and it makes the following more general contributions: (1) we identify some previously ignored features of search spaces that can lead to GA failure; (2) we demonstrate that there are a number of different, independent factors that contribute to the difficulty of search for a GA; and (3) we conclude that any successful research effort into the theory of GA performance must take into account this multiplicity rather than concentrating on only one factor (e.g., deception; see Goldberg, 1989b; Liepins & Vose, 1990; Whitley, 1991; Das & Whitley, 1991).

While this article deals with learning real-valued functions on bit strings, the discussion and results presented here are relevant to researchers using GAs in other contexts as well, for a number of reasons. First, it is important for researchers using a particular machine-learning method (here, GAs) to know what analysis tools exist and what use they have. Schemas and Walsh analysis are such tools, and this article provides a review of these aimed at readers with little or no previous knowledge of GA theory. Second, Tanese's results had seemingly surprising implications for the relative effectiveness of partitioned versus traditional GAs and for the relative effectiveness of hillclimbing versus either form of the GA. This article explains these results and in so doing, illuminates the particular aspects of Tanese's applications problems that allowed these implications to hold. Finally, this article makes a number of more general points about the GA related to De Jong's advice for using GAs in the context of machine learning:

> The key point in deciding whether or not use genetic algorithms for a particular problem centers around the question: what is the space to be searched? If that space is well-understood and contains structure that can be exploited by special-purpose search techniques, the use of genetic algorithms is generally computationally less efficient. If the space to be searched is not so well understood and relatively unstructured, and *if an effective GA representation of that space can be developed*, then GAs provide a surprisingly powerful search heuristic for large, complex spaces (De Jong, 1990b, p. 351, italics added).

It is of central importance to understand what constitutes "an effective GA representation." We address this question by discussing some more general implications of Tanese's results for studying how the structure of a search space is related to the expected performance of the GA.

In the following sections we review schema processing in GAs, give an overview of the relationship between schema processing and Walsh polynomials, and describe how Walsh analysis has been used to characterize the difficulty of functions for the GA. We then describe the particular functions Tanese used in her experiments (the *Tanese functions*), and discuss and explain some of her anomalous results. Finally, we discuss the more general question of how to characterize problems in terms of the likelihood of successful GA performance (and the role of *GA-deception* in such characterizations). This article presents details from experiments that were summarized in Forrest and Mitchell (1991).

2. Genetic algorithms and schema processing

The notion of a *schema* is central to understanding how GAs work. Schemas are sets of individuals in the search space, and the GA is thought to work by directing the search towards schemas containing highly fit regions of the search space. The notion of schema processing is important in any GA application, but for the purpose of this article, we will restrict our discussion to searches over bit strings. Much of the discussion given here also applies to more general representations.

In the case of bit strings, a schema can be expressed as a template, defined over the alphabet {0, 1, *}, that describes a pattern of bit strings in the search space $\{0, 1\}^l$ (the set of bit strings of length l). For each of the l bit-positions, the template either specifies the value at that position (1 or 0), or indicates by the symbol * (referred to as *don't care*) that either value is allowed.

For example, consider the two strings A and B:

```
A = 100111
B = 010011.
```

There are eight schemas that describe the patterns these two strings have in common, including: ****11, **0***, **0**1, **0*11.

A bit string x that matches a schema s's pattern is said to be an *instance* of s (sometimes written as $x \epsilon s$); for example, 00 and 10 are both instances of *0. In schemas, 1's and 0's are referred to as *defined bits*; the *order* of a schema is simply the number of defined bits in that schema. The *defining length* of a schema is the distance between the leftmost and rightmost defined bits in the string. For example, the defining length of **0*11 is 3, and the defining length of **0*** is 0.

Schemas can be viewed as defining hyperplanes in the search space $\{0, 1\}^l$, as shown in figure 1. Figure 1 shows four hyperplanes (corresponding to the schemas 0***, 1****, *0***, and *1***). Any point in the space is simultaneously an instance of two of these schemas. For example, the point in the figure is a member of both 1**** and *0*** (and also of 10***).

The fitness of any bit string in the population gives some information about the average fitness of the 2^l different schemas of which it is an instance (where l is the length of the string), so an explicit evaluation of a population of M individual strings is also an implicit evaluation of a much larger number of schemas. That is, at the explicit level the GA searches through populations of bit strings, but we can also view the GA's search as an *implicit* schema sampling process. At the implicit level, feedback from the fitness function, combined with selection and recombination, biases the sampling procedure over time away from those hyperplanes that give negative feedback (low average fitness) and towards those that give positive feedback (high average fitness).

According to the *building blocks* hypothesis (Holland, 1975; Goldberg, 1989c), the GA initially detects biases in low-order schemas (those with a small number of defined bits),

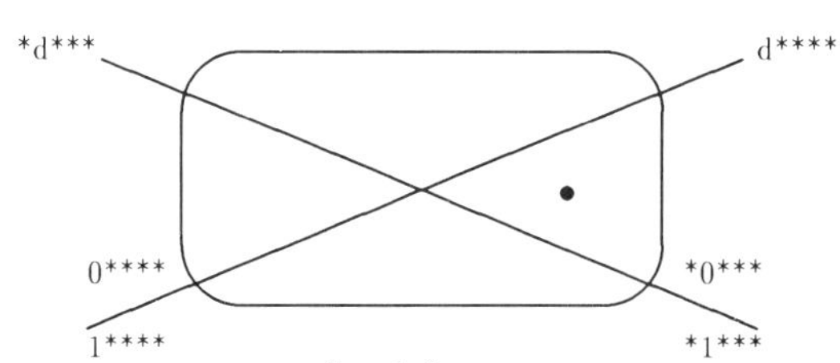

Figure 1. Schemas define hyperplanes in the search space.

and coverages on this part of the search space. Over time, it detects biases in higher-order schemas by combining information from low-order schemas via crossover, and (if this process succeeds) it eventually converges on the part of the search space that is most fit. The building blocks hypothesis states that this process is the source of the GA's power as a search and optimization method. An important theoretical result about GAs is the Schema Theorem (Holland, 1975; Goldberg, 1989c), which guarantees that over time the observed best schemas will receive (in expectation) an exponentially increasing number of samples.

The GA therefore *exploits* biases that it finds by focusing its search more and more narrowly on instances of fit schemas, and it *explores* the search space using the heuristic that higher-order schemas with high average fitness are likely to be built out of lower-order schemas with high average fitness. Reproduction is the exploitation mechanism, and genetic operators—usually crossover and mutation—are the exploration mechanisms. (Holland proposes that mutation is a much less powerful exploration mechanism than crossover, preventing information from being permanently lost from the population (Holland, 1975). According to the Schema Theorem, exploration predominates early on in a run of the GA, but over time, the GA converges more and more rapidly on what it has detected as the most fit schemas, and exploitation becomes the dominant mechanism.

This strong convergence property of the GA is a two-edged sword. On the one hand, the fact that the GA can identify the fittest part of the space very quickly is a powerful property; on the other hand, since the GA always operates on finite-size populations, there is inherently some sampling error in the search, and in some cases the GA can magnify a small sampling error, causing *premature convergence* (Goldberg, 1989c). Also, in some cases strong convergence is inappropriate, for example, in classifier systems (Holland, 1986), in which the GA is trying to evolve a set of co-adapted rules, each one specialized for a specific but different task, rather than a population of similar rules.

As an example of the relevance of schemas to function optimization, consider the function shown in figure 2. The function is defined over integers in the interval [0, 31] (here, $l = 5$), so the x-axis represents the bit string argument (input to the function) and the y-axis

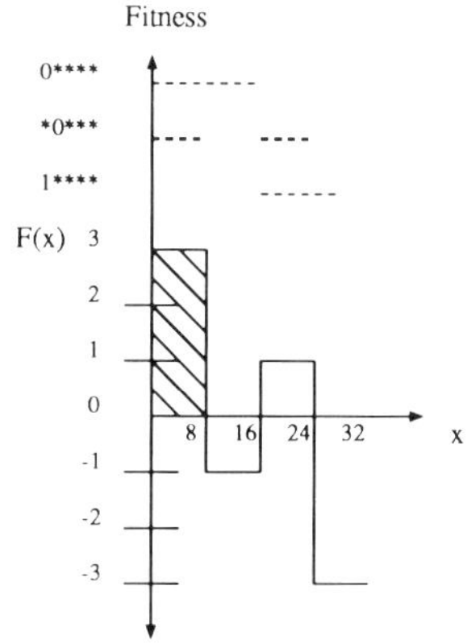

Figure 2. Example function. The solid line indicates the function, and the dashed lines indicate some schemas. The shaded region is the most fit region of the space.

shows the function's value, or fitness. In this example the x value will always be between 0 and 31. For example, the string 10000 would be interpreted as 16, and 01000 would be interpreted as 8.[1] Three schemas are displayed at the top of the plot: the schema 0**** (top dashed line) includes all points less than 16, the schema 1**** (bottom dashed line) includes all points greater than or equal to 16, and the schema *0*** (middle dashed lines) specifies the intervals [0, 8) and [16, 24). Using this example it is easy to see how an individual that was an instance of the schema 0**** could be combined through crossover with an instance of the schema *0*** to yield an instance of 00***, which corresponds to the most fit region of the space (the shaded region in figure 2). That is, 0**** and *0*** are partial solutions.

Schemas induce a partitioning of the search space (Holland, 1988). For example, as seen in figure 1, the partition d**** (where "d" means "defined bit") divides the search space into two halves, corresponding to the schemas 1**** and 0****. That is, the notation d**** represents the partitioning that divides the space into two halves consisting of schemas with a single defined bit in the leftmost position. Similarly, the partition *d*** divides the search space into a different two halves, corresponding to the schemas *1*** and *0***. The partition dd*** represents a division of the space into four quarters, each of which corresponds to a schema with the leftmost two bits defined. Any partitioning of the search space can be written as a string in $\{d, *\}^l$, where the *order* of the partition is the number of defined bits (number of d's). Each partitioning of n defined bits contains 2^n *partition elements*; each partition element corresponds to a schema. Each different partitioning of the search space can be indexed by a unique bit string in which 1's correspond to the partition's defined bits and 0's correspond to the non-defined bits. For example, under this enumeration, the partition d*** . . . * has index j = 1000 . . . 0, and the partition dd*** . . . * has index j = 11000 . . . 0.

3. Walsh-schema analysis

Two goals for a theory of genetic algorithms are (1) to describe in detail how schemas are processed, and (2) to predict the degree to which a given problem will be easy or difficult for the GA. Bethke's dissertation (1980) addressed these issues by applying Walsh functions (Walsh, 1923) to the study of schema processing in GAs. In particular, Bethke developed the *Walsh-schema* transform, in which discrete versions of Walsh functions are used to calculate schema average fitnesses efficiently. He then used this transform to characterize functions as easy or hard for the GA to optimize. Bethke's work was further developed and explicated by Goldberg (1989a, 1989b). In this section we introduce Walsh functions and Walsh polynomials, review how the Walsh schema transform can be used to understand GAs, and sketch Bethke's use of this transform for characterizing different functions. Our discussion is similar to that given by Goldberg (1989a).

3.1. Walsh functions, Walsh decompositions, and Walsh polynomials

Walsh functions are a complete orthogonal set of basis functions that induce transforms similar to Fourier transforms. However, Walsh functions differ from other bases (e.g., trigonometric functions or complex exponentials) in that they have only two values, +1 and −1. Bethke demonstrated how to use these basis functions to construct functions with varying degrees of difficulty for the GA. In order to do this, Bethke used a discrete version of Walsh's original continuous functions. These functions form an orthogonal basis for real-valued functions defined on $\{0, 1\}^l$.

The discrete Walsh functions map bit strings x into $\{1, -1\}$. Each Walsh function is associated with a particular partitioning of the search space. The Walsh function corresponding to the jth partition (where, as described above, the index j is a bit string) is defined as follows (Bethke, 1980; Tanese, 1989a):

$$\psi_j(x) = \begin{cases} 1 & \text{if } x \wedge j \text{ has even parity (i.e., an even number of 1's)} \\ -1 & \text{otherwise.} \end{cases}$$

Here, $\wedge$ stands for bitwise AND. Notice that $\psi_j(x)$ has the property that the only bits in x that contribute to its value are those that correspond to 1's in j.

Plots of the four Walsh functions defined on two bits are given in figure 3.

Since the Walsh functions form a basis set, any function $F(x)$ defined on $\{0, 1\}^l$ can be written as a linear combination of Walsh functions:

$$F(x) = \sum_{j=0}^{2^l-1} \omega_j \psi_j(x)$$

where x is a bit string, l is its length, and each ω_j is a real-valued coefficient called a *Walsh coefficient*. For example, the function shown in figure 2 can be written as

$$F(x) = 2\psi_{01000}(x) + \psi_{10000}(x).$$

The Walsh coefficients ω_j of a given function F can be obtained via the *Walsh transform*, which is similar to a Fourier transform. The representation of a function F in terms of a set of Walsh coefficients ω_j is called F's *Walsh decomposition* or *Walsh polynomial*. In this and the next subsection, we will explain how the Walsh transform works and discuss the close relationship between Walsh analysis and schemas.

As a simple example of the Walsh transform, consider the function $F(x) = x^2$, where x is a two-bit string. The space of two-bit strings can be partitioned into sets of schemas in four different ways, as illustrated in figure 4.

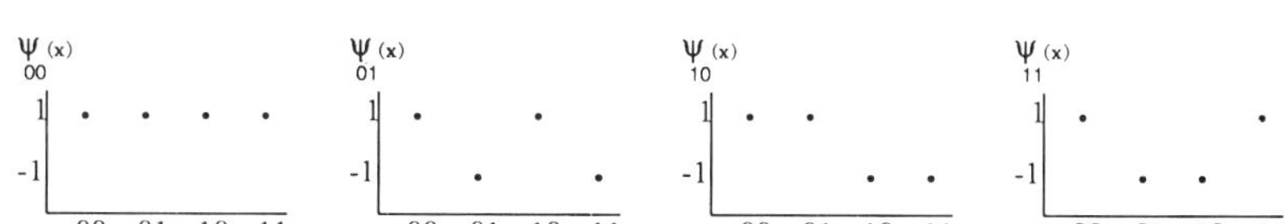

Figure 3. Plots of the four Walsh functions defined on two bits.

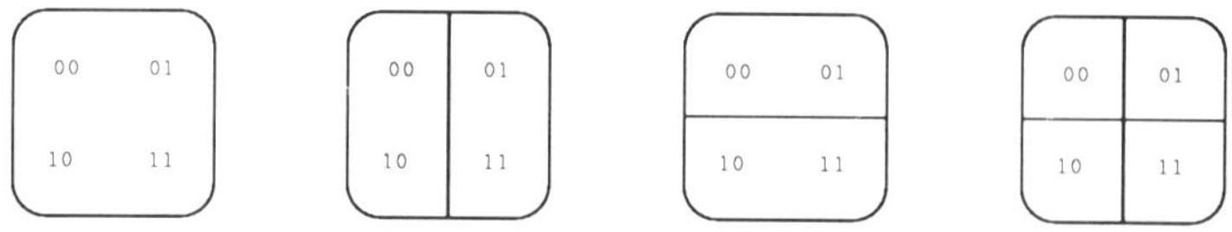

A. Partition = **, j = 00 B. Partition = *d, j = 01 C. Partition = d*, j = 10 D. Partition = dd, j = 11

Figure 4. Four different partitionings of the space of two-bit strings.

The Walsh transform works by transforming $F(x)$ into the summed series of Walsh terms $F(x) = \Sigma_{j=0}^{2^l-1} \omega_j \psi_j(x)$, in which increasingly longer partial sums provide progressively better estimates of the value of $F(x)$. The terms in the sum are obtained from the average values of F in progressively smaller partition elements.

In this example we will use Walsh analysis to get better and better estimates for $F(11)$ (= 9).

Consider first the average value of F on the entire space, which is the same as the average fitness $u(**)$ of the schema ** in the partition $j = 00$ (part A of figure 4):

$$u(**) = \bar{F} = (F(00) + F(01) + F(10) + F(11))/4 = 14/4.$$

Let $\omega_{00} = u(**) = \bar{F}$. This could be said to be a "zeroth-order" estimate of $F(11)$ (or of $F(x)$ for any x).

Now to get a better estimate for $F(11)$, some corrections need to be made to the zeroth-order estimate. One way to do this is to look at the average value of F in a smaller partition element containing $F(11)$—say, *1 (the right-hand element shown in part B of figure 4). The average value of the schema *1 is

$$u(*1) = \omega_{00} - correction_{*1};$$

that is, it is equal to the average of the entire space minus the deviation of $u(*1)$ from the global average. Likewise, the average value of the complement schema *0 is

$$u(*0) = \omega_{00} + correction_{*1},$$

since $u(*1) + u(*0) = 2u(**) = 2\omega_{00}$. (The assignment of + or − to $correction_{*1}$ here is arbitrary; it could have been reversed.) The magnitude of the correction is the same for both schemas (*1 and *0) in partition *d. Call this magnitude ω_{01}. A better estimate for F(11) is then $\omega_{00} - \omega_{01}$.

The same thing can be done for the other order-1 schema containing 11: 1*. Let ω_{10} be the deviation of the average value in d* from the global average. Then,

$$u(1*) = \omega_{00} - \omega_{10}.$$

An even better estimate for $F(11)$ is $\omega_{00} - \omega_{01} - \omega_{10}$. This a first-order estimate (based on one-bit schemas). The two correction terms are independent of each other, since they correct for differences in average values of schemas defined on different bits, so we subtract them both. If the function were linear, this would give the exact value of $F(11)$. (In some sense, this is what it *means* for a function defined on bit strings to be linear.)

However, since $F(x) = x^2$ is nonlinear, one additional correction needs to be made to account for the difference between this estimate and the average of the order-2 schema, the string 11 itself:

$$F(11) = \omega_{00} - \omega_{01} - \omega_{10} + correction_{11}.$$

The magnitude of the order-2 correction term is the same for each $F(x)$. This can be shown as follows. We know that

$$F(11) = \omega_{00} - \omega_{01} - \omega_{10} + correction_{11},$$

and by a similar analysis,

$$F(10) = \omega_{00} + \omega_{01} - \omega_{10} + correction_{10}.$$

Adding both sides of these two equations, we get

$$F(11) + F(10) = 2\omega_{00} - 2\omega_{10} + correction_{11} + correction_{10}.$$

But $F(11) + F(10) = 2u(1*)$ (by definition of $u(1*)$), so we have

$$F(11) + F(10) = 2u(1*) = 2\omega_{00} - 2\omega_{10},$$

since, as we discussed above, $u(1*) = \omega_{00} - \omega_{10}$. Thus, $correction_{11} = -correction_{10}$.

Similarly,

$$F(01) = \omega_{00} - \omega_{01} + \omega_{10} + correction_{01},$$

so,

$$F(11) + F(01) = 2\omega_{00} - 2\omega_{01} + correction_{11} + correction_{01},$$

and since

$$F(11) + F(01) = 2u(*1) = 2\omega_{00} - 2\omega_{01},$$

we have $correction_{11} = -correction_{01}$.

Finally,

$$F(00) = \omega_{00} + \omega_{01} + \omega_{10} + correction_{00},$$

so,

$$F(00) + F(01) = 2\omega_{00} + 2\omega_{10} + correction_{11} + correction_{01},$$

and since

$$F(00) + F(01) = 2u(0^*) = 2\omega_{00} + 2\omega_{10},$$

we have $correction_{00} = -correction_{01}$. Thus the magnitudes of the second-order correction terms are all equal. Call this common magnitude ω_{11}.

We have shown that, for this simple function, each partition j' has a single $\omega_{j'}$ associated with it, representing the deviation of the real average fitness of each schema in partition j' from the estimates given by the combinations of lower-order ω_j's. The magnitude of this deviation is the same for all schemas in partition j'. This was easy to see for the first-order partitions, and we showed that it is also true for the second-order partitions (the highest-order partitions in our simple function). In general, for any partition j, the average fitnesses of schemas are mutually constrained in ways similar to those shown above, and the uniqueness of ω_j can be similarly demonstrated for js of any order.

Table 1 gives the exact Walsh decomposition for each $F(x)$.

We have now shown how function values can be calculated in terms of Walsh coefficients, which represent progressively finer correction terms to lower-order estimates in terms of schema averages. A converse analysis demonstrates how the ω_j's are calculated:

$$\begin{aligned}\omega_{00} &= u(^{**})\\ &= (0 + 1 + 4 + 9)/4\\ &= 14/4\end{aligned}$$

$$\begin{aligned}\omega_{01} &= \omega_{00} - u(^*1)\\ &= (0 + 1 + 4 + 9)/4 - (1 + 9)/2\\ &= (0 - 1 + 4 - 9)/4\\ &= -6/4\end{aligned}$$

$$\begin{aligned}\omega_{10} &= \omega_{00} - u(1^*)\\ &= (0 + 1 + 4 + 9)/4 - (4 + 9)/2\\ &= (0 + 1 - 4 - 9)/4\\ &= -12/4.\end{aligned}$$

$$\begin{aligned}\omega_{11} &= F(11) - \text{first-order estimate}\\ &= F(11) - (\omega_{00} - \omega_{01} - \omega_{10})\\ &= 9 - (14/4 + 6/4 + 12/4)\\ &= 4/4.\end{aligned}$$

Table 1. Expressions for $F(x)$ for each $x \epsilon \{0, 1\}^2$.

$F(00)$	$= \omega_{00} + \omega_{01} + \omega_{10} + \omega_{11}.$
$F(01)$	$= \omega_{00} - \omega_{01} + \omega_{10} - \omega_{11}.$
$F(10)$	$= \omega_{00} + \omega_{01} - \omega_{10} - \omega_{11}.$
$F(11)$	$= \omega_{00} - \omega_{01} - \omega_{10} + \omega_{11}.$

And to check:

$$F(11) = \omega_{00} - \omega_{01} - \omega_{10} + \omega_{11} = 14/4 + 6/4 + 12/4 + 4/4 = 9.$$

In general,

$$\omega_j = \frac{1}{2^l} \sum_{x=0}^{2^l-1} F(x)\psi_j(x).$$

This is the *Walsh transform* (it is derived more formally in Goldberg (1989a)). Once the ω_j's have been determined, F can be calculated as

$$F(x) = \sum_{j=0}^{2^l-1} \omega_j \psi_j(x).$$

This expression is called the *Walsh polynomial* representing $F(x)$.

How does one decide whether or not a deviation term ω_j is added or subtracted in this expression? The answer to this question depends on some conventions: e.g., whether $u(*1)$ is said to be $\omega_{00} - \omega_{01}$ or $\omega_{00} + \omega_{01}$. Once these conventions are decided, they impose constraints on whether higher-order Walsh coefficients will be added or subtracted in the expression for $F(x)$. If x happens to be a member of a schema s whose average deviates in a positive way from the lower-order estimate, then the positive value of the ω_j corresponding to s's partition goes into the sum. All that is needed is a consistent way of assigning these signs, depending on the partition j and what element of j a given bit string x is in. The purpose of the Walsh functions $\psi_j(x)$ is to provide such a consistent way of assigning signs to ω_j's, via bitwise AND and parity. This is not the only possible method; a slightly different method is given by Holland for his *hyperplane transform* (Holland, 1988)—an alternative formulation of the Walsh-schema transform, described in the next section.

3.2. The Walsh-schema transform

There is a close connection betwen the Walsh transform and schemas. The Walsh-schema transform formalizes this connection. In the previous section, we showed that, using Walsh coefficients; we can calculate a function's value on a given argument x using the average fitnesses of schemas of which that x is an instance. An analogous method, proposed by Bethke (1980), can be used to calculate the average fitness $u(s)$ of a given schema s, e.g., *1. Bethke called this method the *Walsh-schema transform*. This transform gives some insight into how schema processing is thought to occur in the GA. It also allowed Bethke to state some conditions under which a function will be easy for the GA to optimize, and allowed him to construct functions that are difficult for the GA because low-order schemas lead the search in the wrong direction.

Formal derivations of the Walsh-schema transform are given by Bethke (1980), Goldberg (1989a), and Tanese (1989a). Here we present the transform informally.

Using the same example as before, the average fitness of the schema *1 is $u(*1) = \omega_{00} - \omega_{01}$; this comes from the definition of ω_{01}. The value of $u(*1)$ does not depend on, say, ω_{10}; it depends only on Walsh coefficients of partitions that either contain *1 or contain a superset of *1 (e.g., ** $\supset$ $*$). In general a partition j is said to *subsume* a schema s if it contains some schema s' such that $s' \supseteq s$. For example, the three-bit schema 10* is subsumed by four partitions: dd*, d**, *d*, and ***, which correspond to the j values 110, 100, 010, and 000, respectively. Notice that j subsumes s if and only if each defined bit in j (i.e., each 1) corresponds to a defined bit in s (i.e., a 0 or a 1, not a *).

The Walsh-schema transform expresses the fitness of a schema s as a sum of progressively higher-order Walsh coefficients ω_j, analogous to the expression of $F(x)$ as a sum of progressively higher-order ω_j's. Just as each ω_j in the expression for $F(x)$ is a correction term for the average fitness of some schema in partition j containing x, each ω_j in the expression for $u(s)$ is a correction term, correcting the estimate given by some lower-order schema that contains s. The difference is that for $F(x)$, all 2^l partition coefficients must be summed (although some of them may be zero). But to calculate $u(s)$, only coefficients of the subsuming partitions ("subsuming coefficients") need to be summed.

The example two-bit function given above was too simple to illustrate these ideas, but an extension to three bits suffices. Let $F(x) = x^2$, but let x be defined over three bits instead of two. The average fitness of the schema *01 is a sum of the coefficients of partitions that contain the schemas ***, **1, *0*, and *01. An easy way to determine the sign of a subsuming coefficient ω_j is to take any instance of s and to compute $\psi_j(x)$. This value will be the same for all $x \epsilon s$, as long as j is a subsuming partition, since all the ones in j are matched with the same bits in any instance of s. For example, the partition **d ($j = 001$) subsumes the schema *11, and $\psi_{001}(x) = -1$ for any $x \epsilon * 11$. Using a similar method to obtain the signs of the other coefficients, we get

$$u(*11) = \omega_{000} - \omega_{001} - \omega_{010} + \omega_{011}.$$

In general,

$$u(s) = \sum_{j:j \text{ subsumes } s} \omega_j \Psi_j(s),$$

where $\Psi_j(s)$ is the value of $\psi_j(x)$ ($= +1$ or -1) for any $x \epsilon s$.

The sum

$$u(*11) = \omega_{000} - \omega_{001} - \omega_{010} + \omega_{011}$$

gives the flavor of how the GA actually goes about estimating $u(*11)$. To review, a population of strings in a GA can be thought of as a number of samples of various schemas, and the GA works by using the fitness of the strings in the population to estimate the fitness of schemas. It exploits fit schemas via reproduction by allocating more samples to them, and it explores new schemas via crossover by combining fit low-order schemas to sample higher-order schemas that will hopefully also be fit. In general there are many more instances of low-order schemas in a given population than high-order schemas (e.g., in a randomly generated population, about half the strings will be instances of 1** ... *, but very few, if any will be instances of 111 ... 1). Thus accurate fitness estimates will be obtained much earlier for low-order schemas than for high-order schemas. The GA's estimate of a given schema s can be thought of as a process of gradual refinement, where the algorithm initially bases its estimate on information about the low-order schemas containing s, and gradually refines this estimate from information about higher- and higher-order schemas containing s. Likewise, the terms in the sum above represent increasing refinements to the estimate of how good the schema *11 is. The term ω_{000} gives the population average (corresponding to the average fitness of the schema ***) and the increasingly higher-order ω_j's in the sum represent higher-order refinements of the estimate of *11's fitness, where the refinements are obtained by summing ω_j's corresponding to higher- and higher-order partitions j containing *11.

Thus, one way of describing the GA's operation on a fitness function F is that it makes progressively deeper estimates of what F's Walsh coefficients are, and biases the search towards partitions j with high-magnitude ω_j's, and to the partition elements (schemas) for which these correction terms are positive.

3.3. The Walsh-schema transform and GA-deceptive functions

Bethke (1980) used Walsh analysis to partially characterize functions that will be easy for the GA to optimize. This characterization comes from two facts about the average fitness of a schema s. First, since $u(s)$ depends only on ω_j's for which j subsumes s, then if the order of j (i.e., the number of 1's in j) exceeds the order of s (i.e., the number of defined bits in s), then ω_j does not affect $u(s)$. For example, ω_{111} does not affect $u(*11)$: *11's two instances 011 and 111 receive opposite-sign contributions from ω_{111}. Second, if the defining length of j (i.e., the distance between the leftmost and rightmost 1's in j) is greater than the defining length of s (i.e., the distance between the leftmost and rightmost defined bits in s), then $u(s)$ does not depend on ω_j. For example, ω_{101} does not affect $u(*11)$—again, since $u(*11)$'s two instances receive opposite-sign contributions from ω_{101}.

Bethke suggested that if the Walsh coefficients of a function decrease rapidly with increasing order and defining length of the j's—that is, the most important coefficients are associated with short, low-order partitions—then the function will be easy for the GA to optimize. In such cases, the location of the global optimum can be determined from the estimated average fitness of low-order, low-defining-length schemas. As we described above, such schemas receive many more samples than higher-order, longer-defining-length schemas: "low order" means that they define larger subsets of the search space and "short defining length" means that they tend to be kept intact under crossover. Thus the GA can estimate their average fitnesses more quickly than those of higher-order, longer-defining-length schemas.

Thus, all else being equal, a function whose Walsh decomposition involves high-order j's with significant coefficients should be harder for the GA to optimize than a function with only lower-order j's, since the GA will have a harder time constructing good estimates of the higher-order schemas belonging to the higher-order partitions j.

Bethke's analysis was not intended as a practical tool for use in deciding whether a given problem will be hard or easy for the GA. As was mentioned earlier, the fitness functions used in many GA applications are not of a form that can be easily expressed as a Walsh polynomial. And even if a function F can be so expressed, a Walsh transform of F requires evaluating F at every point in its argument space (this is also true for the "Fast Walsh Transform" (Goldberg, 1989a)), and is thus an infeasible operation for most fitness functions of interest. It is much more efficient to run the GA on a given function and measure its performance directly than to decompose the function into Walsh coefficients and then determine from those coefficients the likelihood of GA success. However, Walsh analysis can be used as a theoretical tool for understanding the types of properties that can make a problem hard for the GA. For example, Bethke used the Walsh-schema transform to construct functions that mislead the GA, by directly assigning the values of Walsh coefficients in such a way that the average values of low-order schemas give misleading information about the average values of higher-order refinements of those schemas (i.e., higher-order schemas contained by the lower-order schemas). Specifically, Bethke chose coefficients so that some short, low-order schemas had relatively low average fitness, and then chose other coefficients so as to make these low-fitness schemas actually contain the global optimum. Such functions were later termed "deceptive" by Goldberg (1987, 1989b, 1991), who carrried out a number of theoretical studies of such functions. Deception has since been a central focus of theoretical work on GAs. Walsh analysis can be used to construct problems with different degrees and types of deception, and the GA's performance on these problems can be studied empirically. The goal of such research is to learn how deception affects GA performance (and thus why the GA might fail in certain cases), and to learn how to improve the GA or the problem's representation in order to improve performance.

In section 9, we further discuss deception and its relation to the goal of characterizing functions as hard or easy for the GA.

4. Tanese functions

Bethke's method of creating functions by directly assigning the values of Walsh coefficients permitted the straightforward construction of functions that present different degrees of difficulty to the GA. Tanese (1989a, 1989b) also used this method in her dissertation, in which she presented a comparative study of the performance of a number of variants of the GA. In particular, this study compared the variants' performance in optimizing particular classes of Walsh polynomials that were constructed to present various levels of difficulty to the GA.[2]

Tanese compared the behavior of the "traditional" GA with a "partitioned" GA, in which the single large population of the traditional GA was subdivided into a number of smaller subpopulations that independently worked to optimize a given objective function. Tanese also studied a partitioned GA in which individuals in various subpopulations were allowed to "migrate" to other subpopulations during a run (she called this GA "distributed").

For her experiments comparing the performance of GA variants, Tanese wanted to be able to generate automatically a number of classes of fitness functions, with different classes having different levels of difficulty, but with each function in a given class having similar levels of difficulty. She generated different classes of Walsh polynomials as follows. The functions were defined over bit strings of length 32. Each fitness function F was generated by randomly choosing 32 j's, *all of the same order* (e.g., 4). Tanese generated functions only of even order for her experiments. The coefficient ω_j for each of the 32 chosen j was also chosen randomly from the interval (0, 5]. The fitness function consisted of the sum of these 32 terms. Once the 32 j's were chosen, a point x' was chosen randomly to be the global optimum, and the sign of each non-zero ω_j was adjusted so that the fitness of x' would be $\Sigma|\omega_j|$. The *order* of F is defined as the common order of the j's in its terms.

For example,

$$F(x) = \psi_{11110}(x) + 2\psi_{11101}(x) - 3\psi_{11011}(x)$$

is an order-4 function on five-bits (rather than 32) with three terms (rather than 32), with global optimum 6. This Walsh polynomial, like those used by Tanese, has two special properties: (1) all of the non-zero terms in the polynomial are of the same order (Tanese actually used only even-order functions); and (2) there exists a global optimum x' whose fitness receives a positive contribution from each term in the polynomial. Functions with these two properties will hereafter be called *Tanese functions*.

This method of constructing functions had several advantages: (1) it was easy to construct random functions of similar difficulty, since functions of the same order were thought to be of roughly the same difficulty for the GA; (2) functions of different degrees of difficulty could be constructed by varying the order, since low-order functions of this sort should be easier for the GA to optimize than high-order functions; and (3) the global optimum was known, which made it possible to measure the GA's absolute performance.

The results of Tanese's experiments were surprising, and seem to contradict several common expectations about GAs. Specifically, Tanese's results show that for *every* Tanese function she studied—including the low-order ones—the GA performed poorly, and that its performance was often improved when the total population was split up into very small subpopulations. Moreover, Tanese found that the performance of a simple iterated hillclimbing algorithm was significantly better on these functions than both the traditional and partitioned forms of the GA. These results apparently contradict Bethke's analysis, which predicts that the low-order Tanese functions should be relatively easy for the GA to optimize. These results also apparently contradict some other beliefs about the GA—that it will routinely outperform hillclimbing and other gradient descent methods on hard problems such as those with nonlinear interactions (Holland, 1988), and that a population must be of a sufficient size to support effective schema processing (Goldberg, 1985, 1989d). In order to better understand the sources of Tanese's results, we performed a number of additional experiments, which are described in the following sections.

5. Experimental setup

The experiments we report in this article were performed with a similar GA and identical parameter values to those used by Tanese (1989a) (and also in previous work by Forrest, 1985). All of Tanese's experiments used strings of length 32 and populations of 256 individuals. The population was sometimes subdivided into a number of smaller subpopulations. Tanese's algorithm used a *sigma scaling* method, in which the number of expected offspring allocated to each individual is a function of the individual's fitness, the mean fitness of the population, and the standard deviation from the mean. The number of expected offspring given to individual x was $((F(x) - \bar{F})/2\sigma) + 1$, where $\bar{F}$ is the population mean and σ is the standard deviation. Thus an individual of fitness one standard deviation above the population mean was allocated one and a half expected offspring (there was a cutoff at a maximum of five expected offspring). Multipoint crossover was used, with a crossover rate of 0.022 per bit (e.g., for 32-bit strings, there were on average 0.7 crossovers per pair of parents). The crossover rate (per pair) was interpreted as the mean of a Poisson distribution from which the actual number of crosses was determined for each individual. The mutation probability was 0.005 per bit. With the exceptions of sigma scaling and multipoint crossover, Tanese's GA was conventional (proportional selection, complete replacement of the population on every time step, and no creative operators besides crossover and mutation). Tanese ran each of her experiments for 500 generations. For some of our experiments we altered certain parameter values; these exceptions will be noted explicitly.

Tanese conducted experiments on Walsh polynomials of orders 4, 8, 16, and 20. For each experiment she randomly generated 64 functions of a given order, and compared the performance of the *traditional* (single population) GA with a number of *partitioned* GAs in which the population of 256 individuals was subdivided into various numbers of smaller subpopulations. In this article, we discuss results only for functions of order 8, since these were the functions Tanese analyzed in the greatest depth. All of our experiments involved manipulations of parameters in the traditional GA; we did not do any experiments with partitioned GAs. For each of our experiments, we ran the GA once on each of 20 different randomly generated functions, for 200 generations each. Tanese carried each run out to 500 generations, but in each of our runs that used strings of length 32, the population had converged by about generation 100, and the results would not have changed significantly if the run had been extended. The shorter runs were sufficient for determining the comparative effects of the various manipulations we performed on the parameters.

Tanese also compared her GA results with the results of running an iterated hillclimbing algorithm on randomly generated Tanese functions. The iterated hillclimbing algorithm works as follows (Tanese, 1989a).

Repeat the following until a specified number of function evaluations have been performed.

1. Choose a random string x in the search space, and calculate its fitness.
2. Calculate the fitness of every one-bit mutation of x. If an improvement over x's fitness is found, set x equal to the string with the highest fitness.
3. Repeat step 2 until no one-bit mutation yields an improvement. Save this "hilltop" point.
4. Go to step 1.

Finally, return the highest hilltop.

In order to compare the performance of a run of iterated hillclimbing with that of a run of the GA, Tanese ran the above algorithm until the number of function evaluations was equal to that performed by the GA (the population size times the number of generations).

6. An examination of Tanese's results

One of Tanese's most striking results was the poor performance of the GA (in both its traditional and partitioned-population form) on the functions described above and the superior performance of hillclimbing to both forms of the GA.

The results of our replication of some of Tanese's original experiments comparing the traditional GA and hillclimbing are summarized in the first and sixth rows of table 2, under "Original" and "Hill 32" (hillclimbing with 32-bit strings). Our replications confirm Tanese's findings, that, on order-8 functions, neither the GA nor hillclimbing discovers the optimum, but that hillclimbing performs significantly better than the GA in terms of the average maximum fitness found. On average, hillclimbing was able to find a string whose fitness was within about 5% of the maximum, whereas the original GA was able to find a string whose fitness was only within about 12% of the maximum.

In the following subsections, four possible explanations for the GA's poor performance are discussed and evaluated: (1) the choice of performance criteria gave an unfair measure of the GA's performance, (2) crossover is ineffective on these functions because of the lack of lower-order building blocks; (3) the average defining-lengths of the *j*'s are very long, and thus good schemas tend to be broken up by crossover; and (4) the random generation of 32 *j*'s over strings of length 32 results in a large number of correlated positions among the *j*'s, effectively making the functions very difficult.

Table 2. Summary of results of all experiments.

	Times Optimum Found	Average Max Fitness (% opt.)	Average Generation of Max Fitness	Average Max Mean Fitness (% opt.)
Original	0	88.1 (2.9)	31 (33)	85.1 (3.8)
No Xover 32	0	88.4 (2.7)	22 (28)	86.2 (2.6)
Def-Len 10	0	92.3 (2.9)	41 (48)	89.2 (3.0)
Str-Len 128	19	99.97 (0.13)	150 (30)	93.6 (1.3)
No Xover 128	17	99.85 (0.45)	72 (41)	93.9 (0.6)
Hill 32	0	95.4 (1.5)	—	—
Hill 128	20	100.0 (0.0)	—	—

The experiments were all performed running the traditional GA (and in two cases, hillclimbing) on randomly generated order-8 Walsh polynomials. Each result summarizes 20 runs of 200 generations each. The experiments were: (1) *Original* (replicating Tanese's experiments); (2) *No Xover 32* (same as *Original* but with no crossover); (3) *Def-Len 10* (limiting the defining length of partition indices to 10); (4) *Str-Len 128* (incresing the string length to 128 bits); (5) *No Xover 128* (same as *Str-Len 128* but with no crossover); (6) *Hill 32* (hillclimbing on 32 bits); and (7) *Hill 128* (hillclimbing on 128 bits). All runs except the 128-bit runs used strings of length 32. The values given are (1) the number of times the optimum was found; (2) the maximum fitness (percent of optimum) found (averaged over 20 runs); (3) the average generation at which the maximum fitness was first found; and (4) the maximum population mean (% of optimum) during a run (averaged over 20 runs). The numbers in parenthesis are the standard deviations.

6.1. Performance criteria

Tanese used three main performance measures in her experiments, each associated with an average over 64 "trials" (five runs each on 64 randomly generated functions). These measures were: (1) the best overall fitness (averaged over 64 trials); (2) the average fitness in the final generation (averaged over 64 trials); and (3) the best fitness in the final generation (averaged over 64 trials). She found that for the first and last measures, the performance of the partitioned GA was generally superior to that of the traditional GA; these results were generally reversed for the "average final fitness" measure. She also used a summary measure called "success rate": the number of trials on which the global optimum was found at least once out of five times. On the 64 trials (i.e., 320 runs total) on randomly generated order-4 Walsh polynomials, the success rate of the traditional GA was only 3. The most successful partitioned algorithm's success rate was only 15, and the success rate of hillclimbing was 23—almost eight times the success rate of the traditional GA and more than one and a half times the success rate of the best partitioned GA. On 320 runs on randomly generated order-8 polynomials, neither the traditional nor the various partitioned GAs (nor hillclimbing) ever found the optimum, but hillclimbing consistently reached higher fitness values than the GA did.

We examined the possibility that the seemingly poor performance of the GA was due to the strictness of the performance measures used by Tanese. To do this, we compared the GA and hillclimbing using two alternative performance measures that are well known in the GA literature: De Jong's *on-line* and *off-line* performance measures (De Jong, 1975; Goldberg, 1989c). These measures are made in terms of the number of function evaluations an algorithm has performed up to a certain time. The on-line performance at time t is the average value of all fitnesses that have been evaluated by the algorithm up to time t. The off-line performance at time t is the average value, over t evaluation steps, of the best fitness that has been seen up to each evaluation step. For example, if at $t = 5$, five strings have been evaluated yielding fitnesses 10, 20, 8, 4, and 15, the on-line performance would be $10 + 20 + 8 + 4 + 15/5$, and the off-line performance would be $10 + 20 + 20 + 20 + 20/5$.

In figure 5, the on-line and off-line performance values are plotted over 51,200 function evaluations (the equivalent of 200 generations of the GA with a population of 256) for typical runs of both the GA and hillclimbing running on the same order-8 Tanese function. The y-axes give the performance in terms of percent of the function optimum (0%–100%). The on-line and off-line measures were computed every 800 function evalutions. As can be seen, the GA performs significantly better than hillclimbing on on-line performance, but worse on off-line performance (very similar results are obtained on other runs). The on-line result is to be expected: hillclimbing spends most of its time searching through low-fitness single-bit mutations of high-fitness strings, only moving when it finds a higher-fitness mutation. The offline result is closer to traditional optimization measures, and the better performance of hillclimbing on this measure is what is unexpected. In the next several sections, we will propose and evaluate possible explanations for this result.

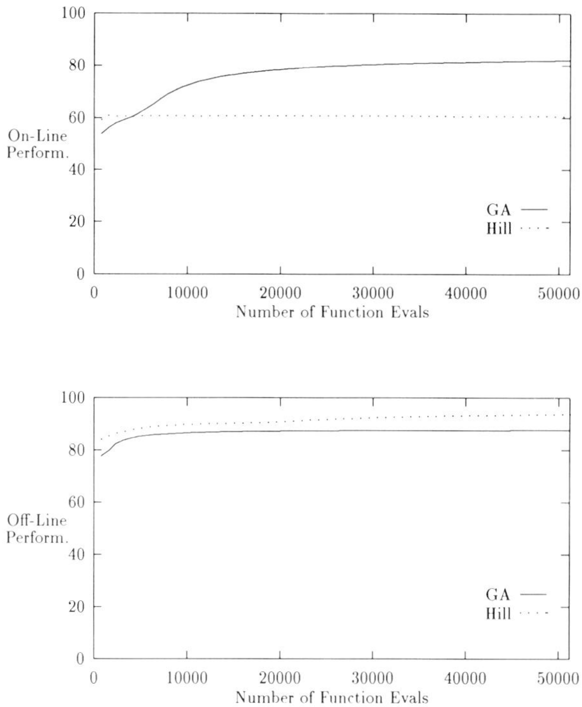

Figure 5. Plots displaying on-line and off-line performance on a typical run of the GA and of hillclimbing on an order-8 Tanese function. The *y*-axes give the performance in terms of percent of the function optimum (0%–100%).

6.2. Is crossover effective on these functions?

As was discussed in section 2, the GA's power is thought to be due to its *implicit* search of schemas in the search space, in which more and more samples are given to fitter schemas, and fit lower-order schemas are combined to produce hopefully fitter higher-order schemas. Thus, to understand the performance of the GA on a fitness function F, we must ask what types of schemas will yield useful information to the GA and to what extent these schemas can be usefully combined by crossover.

The fitness of a string x under a Tanese function F depends on the parity of $x \wedge j$ for each j in the terms of F. Recall that the only terms in F that contribute to a schema s's average fitness are those whose j's subsume s, where j subsumes s implies that each 1 in j corresponds to a defined bit in s. For example, s = *011 is subsumed by j = 0111 (the index for the partition *ddd) but not j = 1011 (the index for the partition d*dd).

To understand this better, consider a Tanese function with only one order-2 term (with $l = 4$):

$$F(x) = 2.0\psi_{1100}(x).$$

A schema s with only one defined bit (order 1) gives no information to the GA about this function, since half the instances of s will yield -2.0 and half will yield $+2.0$. This can be easily checked, for example, with the schema 1***. Likewise, the schema 0*00, which is also not subsumed by $j = 1100$, has the same property. In general, if a term $\omega_j \psi_j(x)$ does not subsume s (i.e., if some 1 in j corresponds to a * in s), then half of s's instances will yield $-\omega_j$ and half will yield $+\omega_j$ for that term, so if a schema s is not subsumed by any terms, then its average fitness is zero. An order-n j cannot subsume a schema of less than order n, so this means that for a Tanese function of order n, no schema of order less than n will give the GA any useful information about what parts of the search space are likely to be more fit. Thus crossover, one of the major strengths of the GA, is not a useful tool for recombining building blocks until schemas that are subsumed by terms in the fitness function have been discovered—such schemas must be of order at least as high as that of the function. This means that the power of crossover—as a means of recombining fit lower-order building blocks—is severely curtailed in these functions.

To verify this empirically, we ran the GA without crossover (i.e., crossover-rate = 0) on 20 randomly generated 32-bit order-8 functions. These results (along with results from all experiments described in this article) are summarized in table 2; they are summarized under the heading *No Xover 32*, and are to be compared with the values under *Original*, giving the results from our replication of Tanese's traditional GA runs on order-8 functions. The GA's performance was not impaired; the maximum fitness discovered and the mean population fitness are not significantly different from the runs in which crossover was used (*Original* in the table).

6.3. Is the GA's poor performance due to long defining lengths of schemas?

In the Tanese functions, an order-n partition index j was constructed by randomly placing n 1's in a string of 32 bits. The expected defining length for such a j can be calculated using equation (4) from Goldberg, Korb, and Deb (1990) (p. 5): $\langle \delta \rangle/(l + 1) = (n - 1)/(n + 1)$, where $\langle \delta \rangle$ is the expected defining length for a string with n 1's in an l-bit string. Thus, the expected defining length for an order-4 Tanese function (with strings of length 32) is approximately 20, and for an order-8 function it is approximately 26, a substantial proportion of the entire string. This last estimate corresponds closely to the empirical data presented in Tanese's thesis (Tanese, 1989a, table 5.1) for one example order-8 Walsh polynomial.

As we pointed out earlier, the only useful schemas are ones that are subsumed by one or more of the terms in the fitness function. The calculation given above implies that such schemas will, like the j's in the function's terms, tend to have long defining lengths. As was discussed in section 3.3, long defining lengths of j's (and thus of useful schemas) can make a function hard for the GA because of the high probability of crossover disruption of useful schemas, since the longer the defining length of a schema, the more likely it is to be broken up under crossover (Holland, 1975; Goldberg, 1989c). To what degree was this problem responsible for the poor performance of the GA on these functions?

To answer this question, we ran the traditional GA with Tanese's parameters on 20 randomly generated order-8 functions, in which the j's were restricted to have a maximum defining length of 10. That is, for each of the 32 j's, a randomly positioned window of 10 contiguous loci was chosen, and all eight 1's were placed randomly inside this window.

The results are summarized in table 2 under *Def-Len 10*. Using the success-rate criterion described above, the performance was identical to Tanese's original results: the traditional GA never found the optimum. Other performance measures made it clear that limiting the defining length improved the GA's performance slightly: the GA was able to find strings with slightly higher fitnesses (in terms of percent of optimum) and slightly higher mean-population fitnesses. We conclude that the contribution of long defining lengths to the GA's overall poor performance on these functions is not significant.

6.4. Is the GA's poor performance due to overlap among significant loci in the partition indices?

Next, we consider the possibility that overlaps among defined positions in the j's (i.e., loci with 1's) were causing correlations among terms in a given fitness function, making the optimization problem effectively higher-order. As a simple example of this, suppose that 0011 and 0110 are two order-2 j's that have non-zero positive coefficients. These j's *overlap* at bit position 2, since they both have value 1 at that locus. There are eight strings x that will cause $\psi_{0011}(x)$ to be positive, corresponding to the schemas **00 and **11. Likewise, there are eight strings that will cause $\psi_{0110}(x)$ to be positive, corresponding to the schemas *00* and *11*. So, in order to find a point that gets a positive value from both $\psi_{0110}(x)$ and $\psi_{0011}(x)$, the GA must discover either the schema *000 or the schema *111. This overlap has the effect that three bits are effectively correlated instead of two, making the problem effectively order-3 instead of order-2. In the case of the Tanese functions, this is a likely source of difficulty; for example, with order-8 functions, where 32 order-8 terms were randomly generated, each 1 in any given j will on average be a member of seven other different j's, and thus the effective linkage will be exceedingly high.

To assess the effect of overlaps, we ran the GA on functions in which the strings were of length 128 rather than 32 (but still order 8), in order to reduce the number of overlaps. With a string length of 128, each defined locus with a 1 would participate on average in only 2 of the 32 j's. As shown in table 2 (under *Str-Len 128*), the GA's performance was remarkably improved. The GA found the optimum 19/20 times (compared with 0/20 for the 32-bit case), and came very close to finding the optimum on the other run. In addition, the mean fitnesses of the population were substantially higher than they were on the 32-bit functions. Of all the experiments we tried, this caused the most dramatic improvement, leading us to conclude that the principle reason the Tanese functions are difficult is because the short strings (32 bits) and relatively high number of terms (32) causes nearly all 32 bits to be correlated, thus creating an order-32 problem. In the non-overlapping case, the order of the problem is much lower, and it is possible for the GA to optimize each term of the function almost independently.

To verify further the ineffectiveness of crossover on the Tanese functions, we ran the GA without crossover on 20 randomly generated 128-bit order-8 functions. The results are summarized under *No Xover 128* in table 2: the performance of the GA was not significantly different from the 128-bit runs in which crossover was used (*Str Len 128*). With both the shorter and longer string-lengths (and thus with both large and small amounts of overlap), whether or not crossover is used does not make a significant difference in the GA's performance on these functions.

The fact that overlap is much higher with 32-bit functions than with 128-bit functions explains the strikingly different dynamics between runs of the GA on the former and latter functions, as shown in figures 6 to 11. These figures are explained below.

A Walsh polynomial F can be thought of as a "constraint satisfaction" problem in which each term in F is a particular constraint. The goal is to find an individual x that satisfies (i.e., receives positive values from) as many terms as possible, and preferably from terms with high coefficients. In a typical run on 32-bit strings, the GA quickly finds its highest-fit individual, typically by generation 30 or so (see the "Average Generation of Max Fitness" column in table 2). Figure 6 plots the percentage of the optimum of the maximum fitness in the population versus time for one typical run. As can be seen, the GA found its highest-fit individual by about generation 40, and this individual's fitness was about 90% of the optimum.

Given an individual x in the population, each term $\omega_j\psi_j(x)$ in the fitness function contributes either a positive or negative value to x's total fitness. If $\omega_j\psi_j(x)$ contributes a positive value, we call x a *positive instance* of that term; otherwise we call x a *negative instance*. Figure 7 shows, for this same run, how the percentages in the population of positive and negative instances of each term vary over time. The y-axis consists of the 32 partition coefficients ω_j of the fitness function, in order of ascending magnitude (recall that these coefficients were each chosen randomly in a range from 0.0 to 5.0, and then their signs were adjusted). The x-axis consists of generation numbers, from 0 to 200. At any point in the plot, the value of the gray-scale represents the percentage of individuals x in the population that are positive instances of the given term at the given time. The darker the gray value, the higher the percentage. As can be seen, very early on (after generation 10), a large subset of the terms have 80%–100% as positive instances, and a smaller subset of terms have only a small number of positive instances.

This result can be explained in terms of the constraint-satisfaction problem. Any individual x with suboptimal fitness receives positive values from only a subset of the terms in the fitness function, leaving the rest of the terms "unsatisfied," that is, yielding negative values on that individual. However, because of the high degree of overlap, the constraint satisfaction

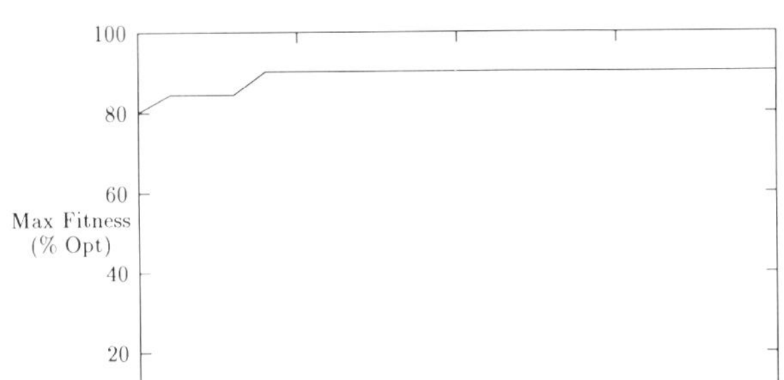

Figure 6. The percentage of the optimum of the maximum fitness plotted over time for a typical run with string legnth 32. For this plot, the maximum fitness was sampled every 10 generations.

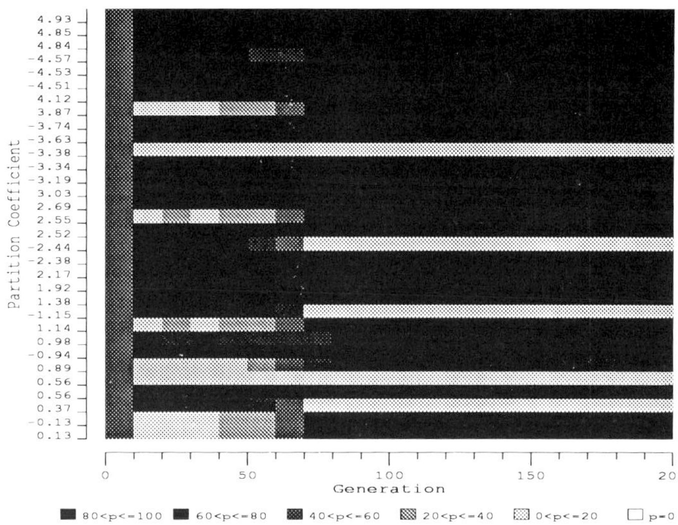

Figure 7. The percent of positive instances for each term in the fitness function plotted over time for the same run with string length 32. The coefficients on the y-axis are in order of increasing magnitude. The darker the gray value, the higher the percentage of positive instances. The values plotted here were sampled every 10 generations.

problem here is very severe, and to discover an individual that receives additional positive values from other terms—without losing too many of the currently held positive values—is very difficult, especially since crossover does not help very much on these functions (see section 6.2). This is particularly true since, once individuals of relatively high fitness are discovered, the diversity of the population falls very quickly. This can be seen in figure 8, which plots the number of distinct individuals in the population over time.[3] As figure 7 indicates, the population very quickly subdivides itself into a small number of mutually exclusive sets: one large set of individuals receiving positive values from one subset of

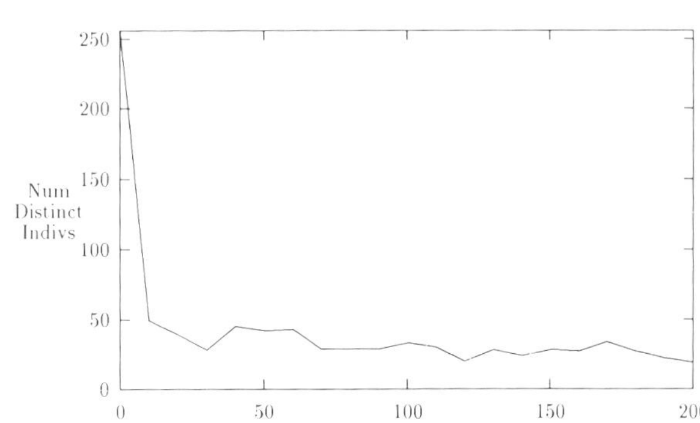

Figure 8. The number of distinct individuals in the population plotted over time for the same run with string length 32. For this plot, the number of distinct individuals was sampled every 10 generations.

the terms in the fitness function, and other, much smaller sets of individuals receiving positive values from the other terms in the fitness function.

To summarize, the GA stays stuck at a local optimum that is discovered early. It is possible that raising crossover and mutation rates (or using other techniques to prevent premature convergence; see Goldberg (1989c)) might improve the GA's performance on such functions, since it would increase the amount of diversity in the population.

On a typical run involving strings of length 128, the situation is very different. Figures 9 to 11 plot the same quantities for one such run. As can be seen in figure 9, the GA tends not to discover its highest-fit individual until late in the run (on average, around generation 150), and, as can be seen in figure 11, the diversity of the population remains relatively high throughout the run. The continuing high diversity of the population is a result of several factors, including the following: since the problem is less constrained, there are many ways in which an individual can achieve relatively high fitness; and since the crossover rate was defined on a per-bit basis, there are on average a greater number of crossovers per chromosome here than in the 32-bit case. But the relative lack of constraints was the major factor, since when crossover was turned off in the 128-bit case, the population diversity remained considerably higher than for the 32-bit run described above, although it was less than in the 128-bit case with crossover turned on. As can be seen in figure 10, in the 128-bit case the population does not segregate into mutually exclusive sets—throughout a typical run, almost all of the terms in the fitness function provide positive values to a significant fraction of the population.

Because of the relative lack of constraints in the 128-bit case, the population does not quickly become locked into a local optimum, and is able both to continue exploring longer than in the 32-bit case and to optimize each term of the fitness function more independently.

In short, these results indicate that the major cause of the GA's difficulty on Tanese's original functions is the high degree of overlap among significant loci in the j's, creating problems of high effective order. A secondary cause of difficulty is the lack of information from lower-order schemas, making crossover relatively ineffective on these functions.

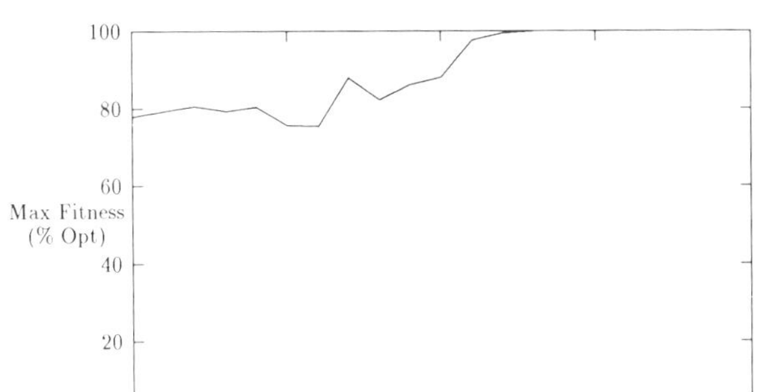

Figure 9. The percentage of the optimum of the maximum fitness plotted over time for a typical run with string length 128. For this plot, the maximum fitness was sampled every 10 generations.

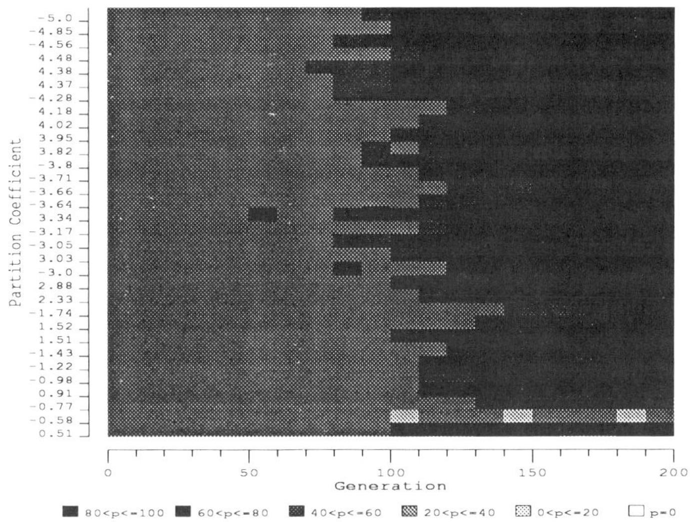

Figure 10. The percent of positive instances for each term in the fitness function plotted over time for the same run with string length 128. The coefficients on the y-axis are in order of increasing magnitude. The darker the gray value, the higher the percentage of positive instances. The values plotted here were sampled every 10 generations.

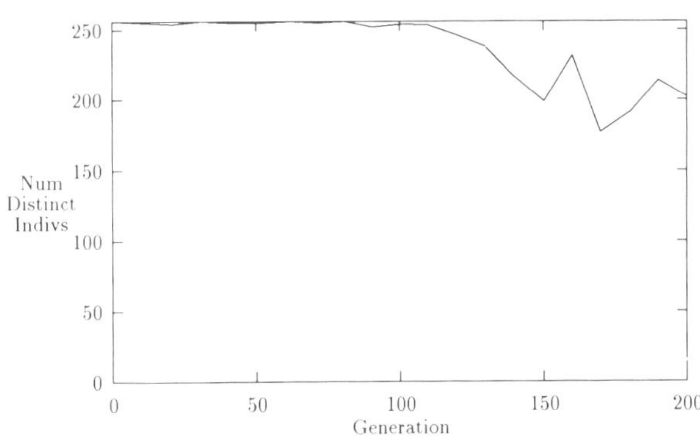

Figure 11. The humber of distinct individuals in the population plotted over time for the same run with string length 128. For this plot, the number of distinct individuals was sampled every 10 generations.

Note that these two factors—overlaps and lack of information from lower-order schemas—do not imply that the Tanese functions are deceptive; a function can have these features and still not be deceptive. The Tanese functions thus provide an important counterexample to a prevailing belief that deception is the crucial feature in characterizing what types of functions are difficult for the GA to optimize (Das & Whitley, 1991). The notion of deception and its relation to the Tanese functions and to GA performance in general is discussed further in section 9.

7. Why does subdividing the population seem to improve performance?

Recall that Tanese compared the behavior of the "traditional" GA with a "partitioned" GA, in which the single large population of the traditional GA was subdivided into a number of smaller subpopulations that independently worked to optimize the given function. Although both the traditional and partitioned GAs performed poorly on the Tanese functions, the results reported by Tanese showed that the performance of most instances of the partitioned GA was better than that of the traditional GA. This was the case even when the population was subdivided into extremely small subpopulations, such as 64 subpopulations of 4 individuals each, and, in some cases, even for 128 subpopulations of 2 individuals each. For functions of order higher than 4, neither the traditional nor partitioned GA's ever found the function optimum, so the difference between the two was measured only in terms of proximity to the optimum of the best fitness discovered.

As Tanese points out, these results run against conventional wisdom about GAs: it has been thought that on difficult problems a large population is needed for processing a sufficient number of schemas (Goldberg, 1985).

Tanese proposes three main reasons for this surprising result.

1. Tanese functions have a large number of local optima, and the GA tends to converge on one. Each of the smaller subpopulations of the partitioned GA will converge earlier than a larger single population, but the subpopulations tend to converge on different optima, and so are able to explore a greater number.
2. After the populations converge, the major adaptive force is mutation. Mutation will be more effective in a smaller population than in a large population, since in a large population, the greater number of mutations will drive up the variance of fitnesses in the population, making it less likely that fit mutations will be able to produce enough offspring to spread in the population (recall that the expected number of offspring is a function of the fitness and the standard deviation). In the smaller population, fewer mutations will occur and the variance will be lower, allowing fit mutations to spread more effectively.
3. Since smaller populations tend to be more homogeneous (for the reasons given above), fit mutations are less likely to be destroyed through crossover.

Explanation 1 seems correct, especially in light of the dynamics that were discussed in section 6.4. It seems that the traditional GA quickly finds a single local optimum that satisfies a subset of terms in the Walsh polynomial, and cannot go beyond that point. Since the number of different chromosomes in the population falls so quickly, the traditional GA does not use the potential for diversity that its large population offers. This explains why a number of small populations running in parallel would likely results in at least one way of satisfying the constraints that was superior to local optimum found by a single large-population GA.

Explanation 2 relies on the assumption that large populations tend to be more diverse and have higher standard deviations than smaller ones. But as was seen in figure 8, the diversity of the large population with 32-bit strings falls very quickly, so this assumption seems incorrect. We believe that the main factor keeping successful mutations at bay is the degree to which the problem is constrained: once a local optimum is settled upon, it is very unlikely that a single mutation (or a small set of mutations) will yield a fitter individual.

A similar point could be made with respect to explanation 3. Given the homogeneity that results even with a large population, it seems likely that this is not a significant factor in the relative poor performance of the traditional GA.

It is important to point out that the effects discussed here come about because of the special characteristics of the fitness functions being used—in particular, the fact that the single-order functions prevent the GA from exploiting information from lower-order building blocks through crossover. For functions in which lower-order building blocks do yield useful information, these effects might not be seen. The results and analysis for these functions cannot be assumed to be applicable to all instances of Walsh polynomials.

8. Why does hillclimbing outperform the GA on these functions?

The discussions in the previous sections allow us to understand why hillclimbing outperforms the GA on the Tanese functions. Since the nature of the Tanese functions precluded the GA from using information from lower-order building blocks via crossover, the genetic operators of crossover and mutation served mainly as a means of generating random variation—in effect, an inefficient form of hillclimbing. This is the reason for the greater success of hillclimbing on these functions. This is shown further by the results of running hillclimbing on functions with 128-bit strings, summarized under *Hill 128* in table 2. On these easier functions (with less overlap than the 32-bit string functions), hillclimbing still beat the GA, funding the optimum 20 out of 20 times versus 17 out of 20 times for the GA.

9. Are the Tanese functions deceptive?

None of the explanations given so far for Tanese's anomalous results relies on the Tanese functions being deceptive. However, the poor performance of the GA led Goldberg to propose that "deception or partial deception" was lurking in the Tanese functions (Goldberg, 1991). In this section we examine the question of whether or not it is possible for a Tanese functions to be deceptive. This examination has led to some general concerns about what the term "deception" actually means or should mean, and what its role is in understanding GAs.

9.1. Definitions of deception

A number of different definitions of deception as well as types of deception have been proposed in the GA literature (e.g., see Bethke, 1980; Goldberg, 1987; Liepins & Vose, 1990; Whitley, 1991). Although thre is no generally accepted definition of the term "deception," it is generally agreed that the notion of deception involves situations in which lower-order schemas give misleading information to the GA about the probable average fitness of higher-order refinements (higher-order schemas contained by the lower-order schemas). For the purpose of this section, we will use the definitions of different types of deception proposed by Whitley (1991). These definitions are summarized below. These definitions all use the notion of competition among schemas in a partition. For example, in a four-bit problem, the order-2 partition *dd* contains four schemas: *00*, *01*, *10*, and *11*. The *winner* of this "hyperplane competition at order-2" (or, equivalently, the *winner of this partition*) is the schema with the highest average fitness. The partition *dd* is said to *subsume* the partition *ddd, which contains eight schemas, each a subset of one of the schemas in *dd*. This notion of a partition subsuming another partition is analogous to the notion of a partition subsuming a schema.

The following definitions are adopted from Whitley (1991) (which, as is stated in that paper, are consistent with most other proposed definitions of deception in the literature).

Definition: A problem *contains deception* if the winner of some partition has a bit pattern that is different from the bit pattern of the winner of a higher-order subsumed partition.

Definition: A problem is *fully deceptive* at order N if, given an order-N partition *P*, the winners of *all* subsuming lower-order partitions lead toward a *deceptive attractor*, a hyperplane in *P* that is different from the winner of *P*.

Definition: A problem is *consistently deceptive* at order N if, given a order-N partition *P*, the various winners of all subsuming lower-order partitions lead toward hyperplanes that are different from the winner of *P* (though not necessarily leading towards the same deceptive attractor).

It should be noted that, strictly speaking, deception is a property of a particular *representation* of a problem rather than of the problem itself. In principle, a deceptive representation could be transformed into a non-deceptive one, but in practice it is usually an intractable problem to find the appropriate transformation.

9.2. Tanese functions and deception

The first question we want to address is, could the functions that Tanese used possibly be deceptive? As it turns out, the answer is not clear, since the definitions given above neglect a key aspect of these functions: every partition of the search space contains at least two winning (maximal) schemas. This is because for j's of even order, $\psi_j(x) = \psi_j(\bar{x})$. But the definitions given above all specify that there is a single winner ("the winner") of each partition.

The definition of *contains deception* could be modified as follows:

Definition: A problem *contains deception* if *one of the winners* of some hyperplane competition has a bit pattern that is different from the bit pattern of *each of the winners* of a higher-order subsumed partition.

If this definition were adopted, then it would be possible for a even-order Tanese function to contain deception. The following is a simple example of such a function (where the bit strings are of length 5 and the order of the j's is 4):

$$F(x) = \psi_{11110}(x) + 2\psi_{11101}(x) - 3\psi_{11011}(x).$$

$F(x)$ meets the two requirements of a Tanese function: all the non-zero terms are of the same order (4), and there is a point x' whose fitness receives a positive contribution from each term in $F(x)$. One such point is $x' = 10100$, with $F(x') = 6$.

To show that this function contains deception according to the modified definition, consider the partition P = dddd*. One winner of P is the schema s_0 = 1111*. This can be seen as follows. As was discussed in section 6.2, the only terms in F that contribute to a schema s's average fitness are those whose j's *subsume s*, where j *subsumes s* implies that each 1 in j corresponds to a defined bit in s. For example, s_0 is subsumed only by the first term of F. A term $\omega_j\psi_j(x)$ that does not subsume s will contribute 0 to s's average fitness, since, because a Walsh function's value depends on parity, half the instances of s will make this term positive and half will make it negative. This can be easily checked for s_0 with respect to the second two terms of F. The average fitness of s_0 is 1, which is the highest possible average fitness for schemas in P, since schemas in P are subsumed only by the first term of F. Thus, s_0 is a winner in P.

However, consider the subsumed partition P' = ddddd. According to the modified definition, if the problem does not contain deception, then either 11111 or 11110 should be a winner in P'. But the fitness of 11111 is 0, and the fitness of 11110 is 2, whereas the fitness of $x' = 10100$ (a winner in P') is 6. Thus, this function contains deception according to the modified definition.

It can be proved, however, that for any Tanese function, every partition contains *some* winner that does not lead the GA astray. This is stated formally by the following theorem:

Theorem 1: Let P be any partition of the search space defined by a Tanese function F. Then there exists a schema s in P such that s is a winner in P and, for any P' subsumed by P, there exists a winning schema s' in P' such that $s' \subset s$.

Proof: Let x' be a global optimum of F. Let s be the schema in P whose defined bits are the same as the corresponding bits in x'. The average fitness of s depends only on the terms in F that subsume s. Call the set of such terms SUBSUMES(s). Since x' is a global optimum of a Tanese function, it receives positive values from all terms in F, so it receives positive values from all terms in SUBSUMES(s). But the value of a term $\omega_j\psi_j(x)$ in SUBSUMES(s) on x' depends only on the bits in x' corresponding to 1's in j. Since SUBSUMES(s) is the set of terms that subsume s, these 1's appear only in positions corresponding to defined bits in s (which are the same as x's defined bits), so every instance of s also receives positive values from all the terms in SUBSUMES(s). Thus s is a winner in P. (Note that if SUBSUMES(s) is empty, then all schemas in P are winners, since they all have average fitness 0.)

Now let P' be a higher-order subsumed partition of P. Let s' be the schema in P' for which $s' \subset s$, and for which the additional defined bits in s' are set so that $x' \subseteq s'$. Then, for the same reason as above, s' is a winner in P'.

We have shown that (1) there can be a winning schema in a given partition that does not contain any winning schema in a higher-order subsumed partition, but (2) there is always at least one winning schema in every partition that contains a winning schema in every higher-order subsumed partition (and thus contains a global optimum).

It is unclear from Whitley's definition whether or not this situation should be called "deception." Theorem 1 implies that the Tanese functions cannot be fully or consistently deceptive, but it seems to be a matter of definition whether or not they can contain deception. Certainly the property illustrated by the example "deceptive" function given above might cause difficulty for the GA, but the degree of difficulty might be different from a function containing deception that strictly conforms to Whitley's definition. It is interesting to note that Wilson (1991) defines a problem to be "GA-easy" if "the highest fitness schema in *every* complete set of competing schemata contains the optimum." Theorem 1 could be interpreted to imply that the Tanese functions are GA-easy, since there is *a* highest-fitness schema in each partition that contains the optimum. Of course, as Tanese demonstrated, many Tanese functions were far from easy for the GA.

As with the notion of "GA-easy," most definitions of deception assume that there will always be a single winner in every partition. However, it cannot be expcected that for every problem posed to the GA, every partition in the search space will contain a unique winning schema. Thus it is of considerable importance, in formulating a definition of deception, to take into account situations such as the multiple partition-winners in the Tanese functions. While the Tanese functions are a particularly extreme example of this, multi-modality of this kind certainly exists in other more natural functions, and we need to be clear about what deception means in these circumstances. Definitions of deception should at least take into account the difference between the kind of deception found in the example function given above, and the kind specified by Whitley's original definition. If they are both to be called "deception," then it is clear that deception is a broad concept with a number of dimensions, and these various dimensions should be identified.

Note that the example of a "deceptive" Tanese function given above is quite different from the functions that Tanese actually used. It may be that the differences (e.g., 32 terms instead of 3 terms, 32-bit strings instead of 5-bit strings, higher-order j's) place additional constraints on the functions that affect whether or not they can be deceptive. We have shown only that it is *possible* to construct a Tanese function (obeying the two criteria stated earlier) that contains "deception." It is still an open question whether or not the original Tanese functions actually contain the type of deception illustrated in the example. And again it is important to note that the two major factors we have identified as responsible for the GA's difficulty on the Tanese functions—overlaps and lack of information from lower-order schemas—are distinct from the notion of deception. We believe it is these features, and not deceptiveness, that makes Tanese functions hard for the GA to optimize.

9.3. Deception and GA performance

The previous subsection raised some concerns suggested by the Tanese functions about how deception should be defined. We also have some concerns about the role played by the notion of deception in understanding GAs. It has been common in the GA literature to characterize problems containing deception as "GA-hard" and problems lacking deception as "GA-easy" (e.g., see Goldberg & Bridges, 1988; Liepins & Vose, 1990; Wilson, 1991), but these are unfortunate identifications, since they imply that every problem containing deception will be hard for the GA and every problem lacking deception will be easy for the GA.

However, the GA can easily succeed on many deceptive problems—for example, when the degree of deception is very slight. In addition, Grefenstette and Baker (1989) have questioned some of the assumptions made in applying Walsh analysis to GAs; their results imply that certain types of deception might not cause the difficulty that Walsh analysis would predict. Conversely, non-deceptive problems can be difficult for the GA: for example, Mitchell, Forrest, and Holland's non-deceptive "Royal Road" functions (Mitchell, Forrest, & Holland, 1992; Forrest & Mitchell, 1992) can be constructed to be quite difficult for the GA. Thus, as the terms are currently used, "GA-hard" does not necessarily imply hard for the GA, and "GA-easy" does not necessariy imply easy for the GA. This is clearly a bad use of terminology.[4]

Thus, deception is not the only factor determining how hard a problem will be for a GA, and it is not even clear what the relation is between the degree of deception in a problem and the problem's difficulty for the GA. There are a number of factors that can make a problem easy or hard, including those we have described in this article: the presence of multiple conflicting solutions or partial solutions, the degree of overlap, and the *amount* of information from lower-order building blocks (as opposed to whether or not this information is misleading). Other such factors include sample error (Grefenstette & Baker, 1989), the number of local optima in the landscape (Schaffer et al., 1989), and the relative differences in fitness between disjoint desirable schemas (Mitchell, Forrest, & Holland, 1992). Most efforts at characterizing the ease or difficulty of problems for the GA have concentrated on deception, but these other factors have to be taken into account as well.

At present, the GA community lacks a coherent theory of the effects on the GA of the various sources of facilitation or difficulty in a given problem; we also lack a complete list of such sources, as well as an understanding of how to define and measure them. Until such a theory is developed, the terms "GA-hard" and "GA-easy" should be defined in terms of the GA's *performance* on a given problem rather than in terms of the *a priori structure* of the problem. A general relation between a given problem's structure and the expected GA performance on that problem is still a very open area of research.

Finally, we are concerned about the focus on finding the global optimum that is implicit in many definitions of deception (and thus of "GA-hard" and "GA-easy"). In many applications it is infeasible to expect the GA to find the global optimum (and in many applications the global optimum is not known). Moreover, it could be argued that in general the GA is more suited to finding good solutions quickly than to finding the absolute best solution to a problem. But some researchers define deception solely in terms of hyperplanes leading toward or away from the global optimum (e.g., see Liepins & Vose, 1990). This narrows the concept of deception and makes it less generally useful for characterizing problems according to probable GA success or failure, especially when "success" does not require that the exact optimum point be found. Also, several so-called deceptive problems are essentially multi-peaked functions with a negligible difference between the global optimum and the runner-up; for many applications it would not matter which one the GA found. Should such cases be classified with the stigma of deception? Or even worse, should such cases be relegated to the foreboding class of "GA-hard?" We need to clarify that it *means* for the GA to succeed or fail as well as what causes it to succeed or fail before we can define "GA-hard" and "GA-easy."

10. Conclusions

After examining several possible causes for the GA's poor performance on the Tanese functions, we reach several conclusions:

- Overlap in the defined loci between different j's in a given Tanese function created functions of much higher effective order than the actual order of their Walsh terms. This is the principle reason for the difficulty of the functions.
- The lack of information from lower-order schemas hampered crossover from contributing to the search. This was a secondary cause of the GA's poor performance, and largely accounts for the superior performance of hillclimbing.
- Long defining lengths in the non-zero partitions contributed slightly to the GA's poor performance but were not a major factor. This is related to the ineffectiveness of crossover on these functions. On other problems for which crossover *is* effective, useful schemas with long defining lengths would be more likely to be destroyed by crossover and would thus cause difficulty for the GA.

Tanese's results could be erroneously interpreted to imply that *all* Walsh polynomials are difficult for GAs, and that hillclimbing will generally outperform the GA on such functions. Such a result would be very negative for the GA, since any real-valued function defined on bit strings can be expressed as a Walsh polynomial. However, the experiments and analyses reported in this article suggest that Tanese's results should not be interpreted as a general negative verdict on GAs. The functions she studied are a highly restricted subset of the class of Walsh polynomials and have several peculiar properties that make them difficult to search. Her results could also mistakenly be taken to imply that partitioned GAs with smaller subpopulations will always outperform traditional GAs. This may not be true for functions in which recombination of lower-order building blocks plays a major role in the search.

These results raise the important question of what problems are well suited for the GA, and more importantly, what features of these problems distinguish the GA from other methods. We have shown in this article that the GA's difficulty with the Tanese functions was due to factors other than deception, which has historically been the focus of theoretical work on GAs. The results discussed in this article underscore the fact that there are a number of different, independent factors that can be sources of difficulty for a GA, and we believe that these various factors must be identified and studied as part of the effort to characterize different problems in terms of the likelihood of GA success. The promise of GAs is based on the belief that "discovery and recombination of building blocks, allied with speedup provided by implicit parallelism, provides a powerful tool for learning in complex domains" (Goldberg & Holland, 1988b), but a better understanding of the details of building-block processing and implicit parallelism, and how they are affected by the structure of a given problem, is needed in order to use GAs more effectively.

One clearly important factor lacking in the Tanese functions is the availability of lower-order building blocks that can combine to produce fit higher-order schemas; such hierarchical schema-fitness structure is what makes crossover an effective operator. The "building blocks hypothesis" (Holland, 1975; Goldberg, 1989c) implies that the degree to which a fitness landscape contains such structure will in part determine the likelihood of success for the GA. Another hypothesized contributing factor is the degree to which the fitness landscape contains different "mountainous" regions of high-fitness points that are separated by "deserts" of low-fitness points (Holland, 1989). In order to travel from a low mountainous region to a higher one, a low-fitness desert must be crossed. Point-mutation methods such as hillclimbing can have very high expected waiting times to cross such deserts (if they can cross them at all); the hypothesis is that the GA will be much more likely to cross such deserts quickly via the crossover operation.

The degree to which these factors are present in a given problem may depend to a large extent on the representation chosen for the problem; the role of representation in GA function optimization has been discussed by Liepins and Vose (1990). We are currently studying the behavior of the GA and other search methods on a class of functions in which the degree to which these and other factors are present can be directly varied (Mitchell, Forrest, & Holland, 1992; Forrest & Mitchell, 1992), and we are investigating ways in which the presence of such features can be detected in a given function with a given representation. We believe that this investigation will lead to a better understanding of the classes of problems for which the GA is likely to be a successful search or learning method.

Acknowledgements

This work is a natural follow-up to Reiko Tanese's thesis work. We are grateful for the interesting results that she uncovered and the clarity with which she reported them in her dissertation. We thank John Holland and Quentin Stout for providing many valuable suggestions for this research, Robert Smith for a number of helpful discussions about the notion of GA deception, and Lashon Booker, Greg Huber, Gunar Liepins, and Rick Riolo for their comments on earlier drafts of this article. We also thank Emily Dickinson, Julie Rehmeyer, and Yuval Roth-Tabak for help in creating the figures and graphs in this article. Support for this project was provided by the Santa Fe Institute, Santa Fe, NM (support to both authors); the Center for Nonlinear Studies, Los Alamos National Laboratory, Los Alamos, NM, Associated Western Universities, and National Science Foundation grant IRI-9157644 (support to S. Forrest); and the Michigan Society of Fellows, University of Michigan, Ann Arbor, MI (support to M. Mitchell).

Notes

1. Most numerical optimization applications actually use a Gray-coded representation of numbers for the GA.
2. In her dissertation, Tanese describes experiments involving several fitness functions. In this article we consider only the results with respect to Walsh polynomials.
3. We counted two individuals as being distinct only if they were different at some significant locus—that is, at a locus in which one of the 32 j's had value 1 (otherwise the two individuals would be equivalent as far as the fitness function was concerned). As one would expect, with strings of length 32, every locus was significant, which is not generally the case for strings of length 128.
4. Stewart Wilson (personal communication) has proposed the term "veridical" to replace "GA-easy" in describing the problems lacking deception.

References

Belew, R.K. & Booker, L.B. (Eds.) (1991). *Proceedings of the Fourth International Conference on Genetic Algorithms*. San Mateo, CA: Morgan Kaufmann.

Bergman, A., & Feldman, M.W. (1990). More on selection for and against recombination. *Theoretical Population Biology, 38(1)*, 68-92.

Bethke, A.D. (1980). Genetic algorithms as function optimizers. (Doctoral dissertation, University of Michigan.) *Dissertation Abstracts International, 41(9)*, 3503B.

Das, R., & Whitley, L.D. (1991). The only challenging problems are deceptive: Global search by solving order-1 hyperplanes. In R.K. Belew & L.B. Booker (Eds.), *Proceedings of the Fourth International Conference on Genetic Algorithms*. San Mateo, CA: Morgan Kaufmann.

Davis, L.D. (Ed.) (1987). *Genetic algorithms and simulated annealing*. Research Notes in Artificial Intelligence. Los Altos, CA: Morgan Kauffmann.

Davis, L.D. (Ed.) (1991). *The handbook of genetic algorithms*. New York: Van Nostrand Reinhold.

De Jong, K.A. (1975). *An analysis of the behavior of a class of genetic adaptive systems*. Unpublished doctoral dissertation, University of Michigan, Ann Arbor, MI.

De Jong, K.A. (Ed.) (1990a). Special issue on genetic algorithms. *Machine Learning, 5(4)*.

De Jong, K.A. (1990b). Introduction to the second special issue on genetic algorithms. *Machine Learning, 5(4)*, 351-353.

De Jong, K.A., & Spears, W. (1991). Learning concept classification rules using genetic algorithms. In *Proceedings of the Twelfth International Conference on Artificial Intelligence*, 651-656. San Mateo, CA: Morgan Kaufmann.

Forrest, S. (1985). *Documentation for prisoner's dilemma and norms programs that use the genetic algorithm*. Unpublished report, University of Michigan, Ann Arbor, MI.

Forrest, S., & Mitchell, M. (1991). The performance of genetic algorithms on Walsh polynomials: Some anomalous results and their explanation. In R.K. Belew & L.B. Booker (Eds.), *Proceedings of the Fourth International Conference on Genetic Algorithms*. San Mateo, CA: Morgan Kaufmann.

Forrest, S., & Mitchell, M. (1992). Towards a stronger building-blocks hypothesis: Effects of relative building-block fitness on GA performance. In L.D. Whitley (Ed.), *Foundations of genetic algorithms 2*. San Mateo, CA: Morgan Kaufmann.

Goldberg, D.E. (1985). *Optimal initial population size for binary-coded genetic algorithms* (Technical Report TCGA Report No. 85001). Tuscaloosa, AL: University of Alabama.

Goldberg, D.E. (1987). Simple genetic algorithms and the minimal deceptive problem. In L.D. Davis (ed.), *Genetic algorithms and simulated annealing*. Research Notes in Artificial Intelligence. Los Altos, CA: Morgan Kaufmann.

Goldberg, D.E. (1989a). Genetic algorithms and Walsh functions: Part I, A gentle introduction. *Complex Systems, 3*, 129-152.

Goldberg, D.E. (1989b). Genetic algorithms and Walsh functions: Part II, Deception and its analysis. *Complex Systems, 3*, 153-171.

Goldberg, D.E. (1989c). *Genetic algorithms in search, optimization, and machine learning*. Reading, MA: Addison Wesley.

Goldberg, D.E. (1989d). Sizing populations for serial and parallel genetic algorithms. In J.D. Schaffer (Ed.), *Proceedings of the Third International Conference on Genetic Algorithms*. San Mateo, CA: Morgan Kaufmann.

Goldberg, D.E. (1991). *Construction of high-order deceptive functions using low-order Walsh coefficients* (Technical Report 90002). Urbana, IL: Illinois Genetic Algorithms Laboratory, Department of General Engineering, University of Illinois.

Goldberg, D.E., & Bridges, C.L. (1988). *An analysis of a reordering operator on a GA-hard problem* (Technical Report 88005). Tuscaloosa, AL: The Clearinghouse for Genetic Algorithms, Department of Engineering Mechanics, University of Alabama.

Goldberg, D.E., & Holland, J.H. (Eds.). (1988a). Special issue on genetic algorithms. *Machine Learning, 3(2-3)*.

Goldberg, D.E., & Holland, J.H. (1988b). Introduction to the first special issue on genetic algorithms. *Machine Learning, 3(2-3)*, 95-99.

Goldberg, D.E., Korb, B., & Deb, K. (1990). Messy genetic algorithms. Motivation, analysis, and first results. *Complex Systems, 3*, 493-530.

Grefenstette, J.J. (Ed.). (1985). *Proceedings of the First International Conference on Genetic Algorithms and Their Applications*. Hillsdale, NJ: Lawrence Erlbaum Associates.

Grefenstette, J.J. (Ed.). (1987). *Genetic Algorithms and Their Applications: Proceedings of the Second International Conference on Genetic Algorithms*. Hillsdale, NJ: Lawrence Erlbaum Associates.

Grefenstette, J.J., & Baker, J.E. (1989). How genetic algorithms work: A critical look at implicit parallelism. In J.D. Schaffer (Ed.), *Proceedings of the Third International Conference on Genetic Algorithms*. San Mateo, CA: Morgan Kaufmann.

Holland, J.H. (1975). *Adaptation in natural and artificial systems*. Ann Arbor, MI: University of Michigan Press.

Holland, J.H. (1986). Escaping brittleness: The possibilities of general-purpose learning algorithms applied to parallel rule-based systems. In R.S. Michalski, J.G. Carbonell, & T.M. Mitchell (Eds.), *Machine learning* (Vol. 2). San Mateo, CA: Morgan Kaufmann.

Holland, J.H (1988). The dynamics of searchers directed by genetic algorithms. In Y.C. Lee (Ed.), *Evolution, learning, and cognition*. Teaneck, NJ: World Scientific.

Holland, J.H. (1989). Using classifier systems to study adaptive nonlinear networks. In D.L. Stein (Ed.), *Lectures in the sciences of complexity* (Vol. 1). Reading, MA: Addison-Wesley.

Koza, J.R. (1990). *Genetic programming: A paradigm for genetically breeding populations of computer programs to solve problems* (Technical Report STAN-CS-90-1314). Stanford, CA: Department of Computer Science, Stanford University.

Liepins, G.E., & Vose, M.D. (1990). Representational issues in genetic optimization. *Journal of Experimental and Theoretical Artificial Intelligence*, *2*, 101–115.

Miller, G.F., Todd, P.M., & Hegde, S.U. (1989). Designing neural networks using genetic algorithms. In J.D. Schaffer (Ed.), *Proceedings of the Third International Conference on Genetic Algorithms* (pp. 379–384). San Mateo, CA: Morgan Kaufmann.

Mitchell, M., Forrest, S., & Holland, J.H. (1992). The royal road for genetic algorithms: Fitness landscapes and GA peformance. In *Proceedings of the First European Conference on Artificial Life*. Cambridge, MA: MIT Press/Bradford Books.

Packard, N.H. (1990). A genetic learning algorithm for the analysis of complex data. *Complex Systems*, *4(5)*.

Schaffer, J.D. (ed.). (1989). *Proceedings of the Third International Conference on Genetic Algorithms*. San Mateo, CA: Morgan Kaufmann.

Smith, S.F. (1980). *A learning system based on genetic adaptive algorithms*. Unpublished Ph.D. dissertation, Computer Science Department, University of Pittsburgh, Pittsburgh, PA.

Tanese, R. (1989a). *Distributed genetic algorithms for function optimization*. Unpublished doctoral dissertation, University of Michigan, Ann Arbor, MI.

Tanese, R. (1989b). Distributed genetic algorithms. In J.D. Schaffer (ed.), *Proceedings of the Third International Conference on Genetic Algorithms*. San Mateo, CA: Morgan Kaufmann.

Walsh, J.L. (1923). A closed set of orthogonal functions. *American Journal of Mathematics*, *55*, 5–24.

Whitley, L.D. (1991). Fundamental principles of deception in genetic search. In G. Rawlins (Ed.), *Foundations of genetic algorithms*. San Mateo, Ca: Morgan Kaufmann.

Whitley, L.D., Dominic, S., & Das, R. (1991). Genetic reinforcement learning with multilayer neural networks. In R.K. Belew & L.B. Booker (Eds.), *Proceedings of the Fourth International Conference on Genetic Algorithms* (pp. 562–569). San Mateo, CA: Morgan Kaufmann.

Wilson, S.W. (1991). GA-easy does not imply steepest-ascent optimizable. In R.K. Belew & L.B. Booker (Eds.), *Proceedings of The Fourth International Conference on Genetic Algorithms*. San Mateo, CA: Morgan Kaufmann.

Received October 7, 1991
Accepted January 31, 1992
Final Manuscript July 23, 1992

INDEX